WEATHER SATELLITE HANDBOOK

FIFTH EDITION

DR. RALPH E. TAGGART, WB8DQT

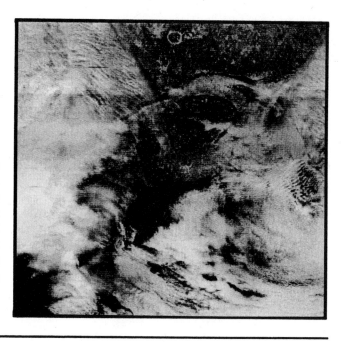

A high-resolution segment of a Russian METEOR visible-light pass in January 1990. The circular feature near top center is Lake Manicouagan, a pair of semicircular lakes in Quebec. The lakes are ice- and snow-covered in this season, and quite prominent. This feature is thought to represent an ancient meteor impact crater. Immediately below the lake is the Gulf of St. Lawerence with Anicosti island to the right.

Foreword

Since the launch of Sputnik in October of 1957, Amateur Radio operators have had a high degree of involvement in space-related activities. Within hours of the announcement of this pioneering spacecraft, the press was turning to the only people in local communities across the world who were equipped to listen for these faint signals—Amateur Radio operators. The local amateur community was already listening in because amateurs possess a unique combination of technical expertise and a sense of pioneering adventure that is the essence of the space program.

Amateurs have continued to be involved in various aspects of the space program. Although amateur-built and operated communications satellites represent the most obvious facet of this involvement, amateurs have been at the forefront of other areas—including the development of ground-station equipment to display weather satellite signals. In 1965, Wendell Anderson, K2RNF, in a pioneering article in *QST*, showed that it was possible to receive and display these fascinating pictures of the earth from space. The international network of weather satellites that has evolved since that time is an expression of multinational involvement and cooperation. Not surprisingly, it has been Amateur Radio operators, more than any other group, who have developed the technology to make this information accessible with modest equipment.

The commitment of the ARRL to publish this latest edition of the *Weather Satellite Handbook* reflects the role of amateurs in enhancing global involvement through the medium of radio and image communications. The history of Amateur Radio has been a continuing story of the advancement of communications technology. Those involved in making weather satellite images accessible to the wider public have been true to this tradition.

David Sumner, K1ZZ
Executive Vice President
February 1994

Preface

The starkly angular spacecraft, about the size of a telephone booth, moves soundlessly along its orbital track over 800 km above the darkened surface of the earth below. A single wing-like solar panel, coldly bathed in starlight, is useless at the moment and the electronic heartbeat of the machine is sustained by internal batteries.

Deep within the spacecraft, an intricate mechanism of motors, lenses, mirrors and solid-state sensors scan the earth below. Some of the sensors are sensitive to visible light and they are as unresponsive as the solar panels to the great black void of the darkened planet. But there are other sensors and these respond to heat variations imperceptible to the eye, building up an image of shorelines, sea ice, frigid clouds, and occasional patches of open ocean in the Arctic night below. Computer circuits process the stream of data from all the sensors and route it to several transmitters connected to spidery antennas on the underside of the spacecraft.

Within minutes the horizon ahead and to the left of the spacecraft begins to brighten, followed by the spectacle of a sunrise as seen from orbit. Details of the earth below now begin to flood the visible light sensors as well. The spacecraft is arcing over extreme northern Canada, Ellesmere Island below is wrapped in an Arctic twilight and the grandeur of ice-locked Hudson Bay becomes visible ahead.

Distorted by the curve of the earth, the Great Lakes seem to roll into view. At that moment, the faint signals from the spacecraft activate a receiver in central Michigan. The warbling tone is noisy at first but strengthens steadily as the spacecraft appears to rise over the northern horizon. As the receiver springs to life, so does a tape recorder, preserving a record of the stream of image data transmitted by the spacecraft. Simultaneously, a picture begins to be painted line by line on a television screen. The hypnotic tone is gradually converted back to an image of light and shadow, reproducing the sweep of Hudson Bay as seen by the spacecraft. As the satellite sweeps southward the images of the Bay are replaced by a stippled pattern characteristic of snow-covered conifer forests. Soon the spacecraft is over the Great Lakes and their characteristic form, easily identified in all seasons, is reproduced on the television monitor. Off to the east, the characteristic outline of Cape Cod, followed by Long Island and the coast of New Jersey and Delaware, make their appearance.

As the spacecraft sweeps over the heartland of the country below, the pattern of cloud and snow cover begins to change and by the time the spacecraft is relaying the image of Florida and the Gulf Coast, all trace of snow is gone and the land below is traced by the patterns of clouds defining a major weather front. The spacecraft continues its ceaseless scanning of the earth below but, as it passes south of Cuba, with South America looming ahead, the Great Lakes, again distorted by the curvature of the earth, seem to fade into the horizon behind it, silencing the receiver and stopping the tape recorder. Despite the fact that no one is present, a computer senses the change in the receiver output and, checking its internal clock, switches the operating frequency to await another spacecraft due over the southern horizon in precisely 37 minutes . . . .

While this description might fit a government installation built to track destructive storms and improve the reliability of weather forecasts, it actually describes a system in my own basement. It, and thousands of others like it, was built solely to gratify my own interest in watching the ever-changing panorama of the earth as viewed from space. This *Handbook*, now in its fifth edition, is designed to introduce you to that same fascinating activity.

The *Weather Satellite Handbook* has changed quite a bit since its first publication in 1976. These changes have been driven by the increasing sophistication of the satellites themselves as well as by the steady march of electronics technology that has made it even easier to watch these elusive images. The fourth edition made the jump to all-digital display technology, featuring a stand-alone scan converter that could communicate with a computer. With the steady drop in the cost of computer hardware, performance continues to escalate. This time around we will be relying entirely on computers to handle our image display chores. While you will end up with an extremely sophisticated display system, the interface to accomplish this is the simplest display project in the history of the *Handbook*. It covers both satellite image formats *and* the ability to process weather charts and satellite photos transmitted on shortwaves!

The goal of the first edition was to introduce as many people as possible to a hobby that is a unique blend of electronics, meteorology, earth science, astronautics, and, now, computer science. It is an activity that has captivated me—to the occasional despair of my family as activities are planned around the orbital timetable of unseen satellites! Earlier editions have served to introduce thousands to that same fascination. If this new edition can do the same for you, it will fulfill its purpose.

Mason, Michigan
February 1994

Acknowledgments

With each edition, the *Weather Satellite Handbook* grows steadily larger. In doing so, it becomes an ever-more-complex project in terms of the design work, the writing, the editorial work, and the actual production. While I am proud to take credit for each new edition, the *Handbook* has reached a level of complexity where it simply wouldn't exist without the input of an ever-larger number of individuals.

I would like to take this opportunity to thank Dr Jeff Wallach, N5ITU, for putting together one of the first comprehensive treatments of the problems and promise inherent in the reception and display of the high-resolution digital imagery from various spacecraft. This is a new feature in this edition that you will find in Chapter 7.

While working on several projects for *QST*, I had the pleasure of doing some software development using the ViewPort VGA SSTV interface, a product of the joint efforts of John Montalbano, KA2PYJ, and Stas Andrzejewski, W6UCM, of A&A Engineering. The flexibility provided by ViewPort served as the inspiration for the WSH Satellite Interface, which appears in Chapter 5. Stas also deserves considerable credit for his contributions to the FM demodulator for this unit. He also did a first-class job of producing many options for the interface, ranging from bare printed-circuit boards, through various levels of kits, to wired and tested units.

Finally, I would be remiss if I failed to recognize the contributions of the energetic and talented editorial and production staff at the American Radio Relay League. When I put the last photograph and computer disk in the Express Mail envelope, I give a silent cheer, pretending that the book is finished. I have done enough books that I know it isn't true, but the good folks at Newington are experts at what they do. In this case, they really "do the book," while I enjoy the illusion that the job is done. My sincere thanks to all of you!

About the Author

Ralph E. Taggart, WB8DQT

Ralph Taggart was born in Charlottesville, Virginia, in 1941. He grew up in a log cabin in the mountains of northern New Jersey, and was licensed as WA2EMC in the late '50s while in high school. He attended Rutgers University and received a BA in Biology in 1963. For two years, Ralph worked as a research assistant at the Boyce Thompson Institute in Yonkers, New York, and spent his spare time working with Lew West (W2PMV) and a small group of northern New Jersey ATV experimenters attempting to establish two-way television communications in the 420-MHz (70-cm) band.

In 1965, Ralph went to Ohio University in Athens, Ohio, to work on a Master's degree in botany, which he received in 1967. The entire rack of home-brew TV cameras, flying-spot scanners, and TV transmitters went along for the duration, sparking the first two-way ATV activity in that part of Ohio. From Athens, Ralph moved on to Michigan State University to work on a PhD in paleobotany (fossil plants) in 1967.

Frustrated with the range limitations of conventional television, he built his first slow-scan television camera and monitor during that first year at MSU. This equipment provided the means for a number of pioneering activities with SSTV through the Michigan State University Radio Club station, W8SH. Among the more notable achievements of this period was the first transmission of full-color images and the first two-way color SSTV contact. It was during this period that Ralph upgraded to Advanced class and received the call sign WB8DQT.

Ralph received his PhD in 1970, and joined the staff at MSU where he now holds an appointment as full Professor in the Department of Botany and Plant Pathology and the Department of Geological Sciences and is curator of fossil plants. In 1972, he co-authored (with Don Miller, W9NTP) the *Slow Scan Television Handbook*, the first comprehensive introduction to SSTV. About the same time, he became interested in applying SSTV display techniques to weather-satellite images. This lead, in 1976, to the publication of the first edition of the *Weather Satellite Handbook*.

Ralph is married, has three daughters, and lives in Mason, Michigan. He is an Elder in the Presbyterian Church, has served on the local Cable TV Advisory Commission, and spent 12 years on the Mason Board of Education (including three years as president). He now serves on the Board of Education for the Ingham Intermediate School District. In addition to his ongoing satellite activities, Ralph is active on UHF ATV and HF SSTV, and enjoys rag-chewing on 80-meter CW. His most recent accomplishment in Amateur Radio was the creation of a new mode, FAX480, for the transmission of VGA-quality images using SSTV hardware. He has been an avid ultralight pilot for twelve years, and for the past three years, he has flown an ultralight gyroplane which he designed.

In addition to his book credits in the area of SSTV and weather satellites, he is the coauthor of two general biology textbooks, numerous research papers and monographs and a chapter on skiflying in a book on ultralight flying techniques. Ralph has written numerous magazine articles that have appeared in *QST*, *73 Magazine*, *Ham Radio Magazine*, *Rotorcraft* and *Kitplanes*.

Dedication

If you are about to become committed to pursuing weather satellites, put aside thoughts of hardware and software. Your greatest asset is an understanding spouse who will tolerate the numerous eccentricities that the hobby demands. I am extremely fortunate to have such an individual in my life: my wife, Alison.

In the earliest days of our marriage, Alison tolerated a student apartment crowded to overflowing with Amateur Radio equipment. Even now she faces a basement filled with ever-more-numerous items of equipment—and nothing ever seems to get thrown away. At times, our life seems dominated by the schedules of unseen satellites, phone calls at all hours from half the world away, radio schedules, equipment tests, and even the writing binge necessary to finish this edition. Through all this, she, and my daughters, Jennifer, Heather, and Molly, have been content merely to mutter at activities that would be cause for institutionalization in a less-tolerant household.

Paul Segal (ex-W9EEA), in the *Amateur's Code*, wisely proclaimed that "The Amateur is Balanced." We may try to achieve that standard but, gripped by the fascination of our hobby, we will almost always fail to some extent. I will promise my bonny Alison that, now that the new edition is finished, I will spend more time with the family. I will promise—and inevitably, I will fail. The marvel is that she will not take that failure as seriously as is her right. For that I am profoundly grateful.

Companion Disk

If you've purchased this book without the optional *Weather Satellite Handbook Program Disk*, you may order it from ARRL Publication Sales. Request ARRL Order No. 4653. The disk is $10, plus $4 for shipping and handling. Prices subject to change.

The IBM-format, 3½-inch, high-density disk includes the following items:

- **WSHDEMO** displays samples of images obtained with the PC-compatible interface described in Chapter 5. The disk includes samples of GOES, NOAA visible, NOAA IR, METEOR and full-Earth-disk images.
- **WSHTRAK** is a satellite tracking program that allows data for up to nine separate spacecraft.
- **FIG10_1.BAS** is a BASIC program for transferring a WSHFAX binary image file to the standard 640×480×16 VGA screen.

Please note: The **WSHFAX** software described in Chapter 5 is *not* included on the disk. It is available separately from A&A Engineering. See Chapter 5 for details.

Updated weather satellite information is available from the WSH Bulletin Board described in Chapter 9 and the Dallas Remote Imaging Group Bulletin Board described in Chapter 5.

For information on any ARRL product, or to place an order, contact

ARRL Publication Sales
225 Main Street
Newington, CT 06111-1494

Voice: 860-594-0200
Fax: 860-594-0303

Contents

Chapter 1

Operational Satellite Systems

INTRODUCTION

There is little doubt that the weather-satellite program of the United States is one of the most tangible benefits of the overall space program. Since the launch of the first operational TIROS satellite in the early '60s, uncounted lives and dollars have been saved as a result of our ability to observe weather phenomena on a global scale. The TIROS satellites of the '60s—primitive by current standards—have been followed by satellites of ever-increasing sophistication. Today, a number of nations are involved in weather observation from space, including the US, Russia, a consortium of European nations that make up the European Space Research Organization (ESRO), and Japan. The People's Republic of China has launched a weather satellite as well.

Most of these operational satellites fulfill their missions with the transmission of very-high-resolution digital images to specially equipped ground stations. It isn't impossible for amateur stations to receive and display such pictures (Figure 1.1), but it does require a good background in microwave techniques and digital electronics (see Chapter 7).

Fortunately, these same satellites transmit lower-resolution analog images that are specifically designed to be easy to receive and display. These images provide the mainstay for amateur weather-satellite activities. Today's operational satellites fall into two general categories: satellites in low, near-polar orbits, and satellites in geostationary orbits.

POLAR-ORBITING SATELLITES

Current operational polar-orbiting satellites include the TIROS/NOAA series operated by the United States and the Meteor/COSMOS satellites of the Commonwealth of Independent States (CIS). These satellites operate in relatively low orbits (in the case of the US satellites, at altitudes of about 600 miles above the earth) and their orbital tracks are such that they come very close to passing over the poles during each revolution of the earth. Essentially, they pass twice over all parts of the earth every 24 hours, once during the day, and again at night.

US TIROS/NOAA Polar Orbiters

The US TIROS/NOAA satellites are very precisely oriented in space in what are known as *sun-synchronous orbits*. This means that during the course of a year, the relationship of the satellite's orbital track to the position of the sun stays relatively constant. Therefore, the satellite passes overhead at about the same solar time each day. Generally, the Russian satellites are not in sun-synchronous orbits, so that the time for optimum passes each day changes in the course of a year.

Although the early weather satellites used TV cameras to obtain their pictures, the delicate vidicon camera tubes they employed are easily damaged, and the performance of a given tube deteriorates steadily with time, leading to marginal images. All of today's operational polar orbiters have replaced the vidicon TV pickup tubes with electromechanical systems known as *scanning radiometers.*

A scanning radiometer is basically a system of lenses, a motor-driven mirror system, and one or more solid-state light sensors that serves as the source of the satellite image data. Basically, the scanning radiometer looks at a very narrow line, equivalent to the horizontal line of a TV picture, at right angles to the satellite's orbital track. The equivalent of the vertical scanning in a TV picture is provided by the motion of the satellite along its orbital track. The scanning system operates continuously, so you can receive a picture that spans the full time that the satellite is within range of your station (Figure 1.2).

Although most of the operations of these satellites can be modified by signals from specially equipped ground stations, picture transmission is essentially automatic and continuous. The pictures that we can receive are usually referred to as *Automatic Picture Transmission,* or APT, imagery.

The image that is produced is mostly a function of

Figure 1.1—A portion of an HRPT image of the central east coast of the United States as received and processed by John Dubois. John has been one of the major forces behind the development of HRPT ground-station equipment that can be duplicated by advanced amateur satellite enthusiasts. It would be possible to write a history of the Civil War using as a background the area encompassed by the southern half of this image. The center is dominated by the great peninsula, containing parts of Delaware and Maryland, bounded to the south by Chesapeake Bay and to the north by Delaware Bay. The west shore of the Chesapeake is dissected by the tidal estuaries of some of the legendary rivers that played a major role in that conflict. To the extreme south is the James River, with the Chickahominy River flowing into it from the north. Further westward is the confluence of the James and the Appomattox Rivers, with the city of Richmond appearing as a small smudge on the James. Immediately north of the James estuary is the York River; north of that is the Rappahannock. The widest of the tidal estuaries, and the next one north, is the Potomac with the city of Washington appearing as a darker smudge where the river abruptly narrows. The city of Baltimore is a slightly larger smudge at the head of the short, but wide, estuary near the northern end of the Chesapeake. The major river flowing into the northern end of the Bay is the Susquehanna. The Delaware flows into the northern end of Delaware Bay, and Philadelphia is clearly visible. In the extreme upper right is the New York metropolitan area, with the wide reach of the lower Hudson River. The many lakes in the Ramapo Mountains area of northern New Jersey and southern New York State are also visible. This particular image was downloaded from the Dallas Remote Imaging Group (DRIG) satellite electronic bulletin board as a Compuserve GIF image, and displayed on an IBM-compatible VGA display system. Chapter 7 describes equipment and techniques for display of HRPT images.

Figure 1.2—A NOAA-11 pass covering most of western North America. Mexico, the US Gulf coast, and Baja California, are the most prominent features in the southern half of the image. There is the typical fog bank off the northern California coast and low fog fills the Central Valley. A major cloud system blankets northern Oregon, Washington, and southern British Columbia, including Vancouver Island. This weather system extends inland and covers much of southern Alberta and Saskatchewan. Manitoba is generally clear, and Lakes Winnepeg and Manitoba are clearly visible as is the western tip of Lake Superior to the southeast. In the extreme upper right is a bit of the coastline of western Hudson Bay. This pass was received by manually tracking the satellite, using a small, 5-element beam. Slightly over 13 minutes of the pass were recorded, and about 12 minutes of that are displayed here. Because the computer displays slightly over 6 minutes of data, the northern and southern halves of the pass were displayed and photographed separately, then joined to prepare this full-pass mosaic. This pass originated in the south, so the original image would have appeared upside down on the display. I have inverted the image here so that north is at the top, making it a bit easier to orient the major features.

...he type of solid-state image sensor being used. Some sensors respond to various portions of the visible-light spectrum; these create images similar, although not necessarily identical, to those you would obtain in a standard photograph (Figures 1.2 and 1.4). The US TIROS/NOAA satellites have two different visible-light sensors, each with slightly different imaging characteristics. In addition to visible-light detectors, the US satellites also have a number of sensors that respond to infrared (IR) or heat radiation. Because each of the

Figure 1.3—The southern half of the same NOAA-11 pass illustrated in Figure 1.2, but showing the IR image data. In the IR format, cold objects (such as high clouds) appear white, while warmer objects are progressively darker. The land masses of Mexico, Baja California, and California are quite dark (hot), contrasting strongly with the cold Pacific Ocean waters to the north. San Francisco Bay is clearly evident on the coast at the upper left and the fog-filled Central Valley shows up well against the warmer surrounding terrain. Note that the offshore fog does not appear in this image. The fog is very low and essentially at sea-surface temperature; thus, it does not contrast with the ocean surface at IR wavelengths.

IR sensors responds to heat and not light, IR cloud pictures can be obtained at night as easily as during the day (Figure 1.3).

The scanning radiometer of a US satellite is actually obtaining a very-high-resolution image at several IR and visible-light wavelengths simultaneously, but this image is transmitted to earth as a wide-bandwidth digital signal that requires very sophisticated equipment to receive and display. Figure 1.1 shows a sample of a portion of this *High Resolution Picture Transmission* (HRPT) format. The high-resolution imaging system is known as a *multispectral scanning radiometer.* Fortunately, an on-board computer is also sampling the high-resolution picture and transmitting those samples in the relatively simple analog APT format that can be received and displayed with comparatively unsophisticated equipment.

Although the various polar-orbiting satellites are quite different from one another, the basic format of their picture signals have a number of common features that make it easy to design a system for handling all of them. Details of the basic video-modulation format are covered in Chapter 4.

The various satellites differ from one another in the rate at which the video lines are transmitted and the available image formats (visible/IR). The TIROS/ NOAA APT transmissions use rates of 120 lines per minute (LPM), and each line is made up of both visible and IR image data. The first half of each line represents the IR data scan, while the second half is the visible-light data. If the image is displayed at 120 LPM, the IR and visible-light images appear side by side. Either the IR or visible-light data can be displayed alone by using a 240-LPM display rate and blanking out every other line. This is what has been done to display the NOAA visible-light image in Figure 1.2 and the IR image of Figure 1.3.

At high latitudes, the quality of visible-light imagery varies with the time of day and the season. The requirement for sun-synchronous orbits restricts the TIROS/NOAA satellites to mid-morning and mid-afternoon passes. During the summer months, the illumination angles are excellent. During the winter, sun angles are lower, and one side of the image is brightly illuminated, while the side farthest from the sun (west in the morning, east in the afternoon) is noticeably darker, particularly for early or late passes.

IR pictures (Figure 1.3) are typically disappointing to those expecting the high contrast typical of visible-light data. In the IR format, warm objects are black, and cold objects are white. The ability to differentiate land, water, and cloud features depends on tempera-

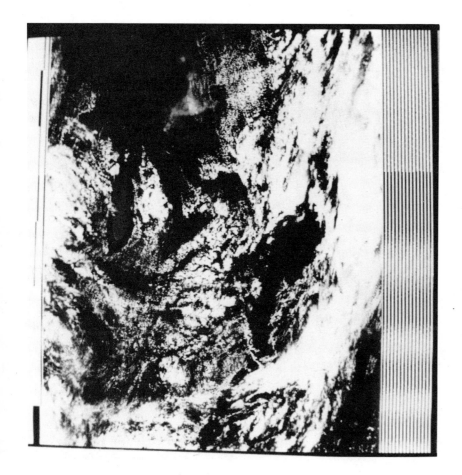

Figure 1.4—An example of typical Meteor warm-weather imagery. Like the passes illustrated in Figure 1.2 and 1.3, this pass originated in the south and ended in the north. This print was inverted to place north at the top. The distinctive 120-LPM Meteor sync-pulse train thus appears at the extreme right, but it actually marks the start of each Meteor image line. As is typical of Meteor images in the summer, cloud details are excellent, but ground features are essentially nil. Michigan and the Great Lakes are just above and to the right of the center of the image. They are evident because the innumerable late-afternoon cumulus-cloud centers do not develop over the cooler lake waters. This particular image was subjected to a considerable amount of logarithmic image contrast in an attempt to resolve ground detail. The only new feature revealed by this processing is a faint sun-glint off the waters of Lake Michigan. With the lake thus defined, you can see how thunderstorm cell development is inhibited near the lake shore and over the lake itself.

ture differences. Ample contrast is available at tropical latitudes and desert areas, but contrast is reduced at higher latitudes. Daylight IR typically has greater contrast than night imagery due to an enhanced thermal gradient, and summer images are better than winter ones. Winter night images at high latitudes may appear almost white, with little or no contrast. IR imaging has great utility because it is not tied to local lighting conditions, but specialized techniques for image enhancement are required to make the most of the small variations when ground temperatures range from cool to cold and certain sensors are in use. Most of the year, it is possible to get useful pictures without special processing, but careful adjustment of video gain is required. The reward is the ability to use evening passes to produce useful imagery, and bring out features (such as the Gulf Stream) that are simply not available in visible-light images.

Russian Polar Orbiters

The workhorse of the Russian operational satellite system is their Meteor satellite series; these transmit visible-light images at a rate of 120 LPM (Figure 1.4). The Meteor satellites do not use sun-synchronous orbits, hence, they pass over in daylight at different

times during their operational lifetime. If a Meteor satellite passes over in early to mid-morning or mid- to late- afternoon, the imagery has the same unequal lighting characteristic of the TIROS/NOAA pictures during the winter months. The Russian satellites tend to have a number of operational satellites in orbit at any time, however, and there is usually one with a useful pass near midday, avoiding the winter-lighting problem.

The US TIROS/NOAA satellites have sensors with a very wide dynamic range, making it possible to differentiate land and water features where lighting is adequate. Meteor sensors tend to compress the black end of the gray scale. Clouds are rendered in great detail, but land/water boundaries are almost indistinguishable without extreme video processing (Figure 1.4). During the winter months at high latitudes, snow and ice cover increases the contrast between land and open water and midday Meteor images can provide excellent results.

For many years, Meteor satellites have provided only visible-light imagery and the satellite transmissions would shut off automatically when light levels dropped below certain thresholds. For a period, the Russians have experimented with a low-resolution IR-imaging system operating at about 20 LPM. The results

Figure 1.5—Approximately 12 minutes of a south-to-north pass of Meteor 3.3 illustrating the Russian 120-LPM IR format. This pass occurred at about 2350Z on a bitter-cold December evening. Most of the Great Lakes are covered by very cold clouds, although enough of the comparatively warm lake waters are visible to help you orient on the major features. The Carolina coast, Delaware, New Jersey, Long Island, and Cape Cod are warmer than the interior, but still contrast against the warmer ocean water. A major cyclonic storm system covers the Canadian Maritimes, but that is not new to those folks! No image enhancement—other than the pixel complementation discussed in the text—was used for this image. It appears that this format has excellent potential for stations that lack the facility to perform image enhancement on winter IR-image data. The familiar Meteor sync-pulse train is off to the right because the mosaic is inverted to place north at the top.

were far from exciting, but it was encouraging that IR experimentation was under way. Late in 1989, the Russians launched a new satellite, Meteor 3/3. This satellite has a 120-LPM IR-imaging system that is currently providing excellent night IR coverage (Figure 1.5). One anomaly of this format is that cold areas (such as clouds) reproduce as black, while warm areas are white. Digital pixel complementation was used to convert the original data in Figure 1.5 to the more familiar cold = white/warm = black format. (This proc-

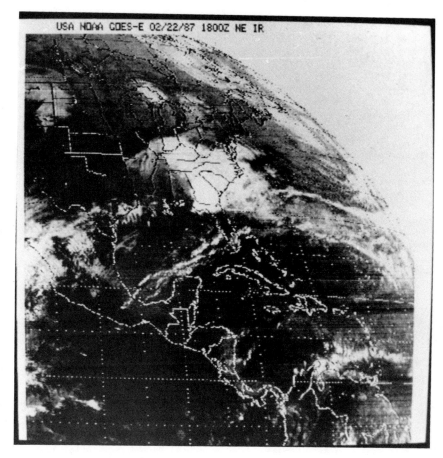

Figure 1.6—A NE IR quadrant, imaged by the GOES East satellite and received via the GOES Central satellite. High, cold clouds are white in the IR format, as is the view of space beyond the limb of the earth. Ground computers insert the political and geographic reference data because some ground features may be obscure at IR wavelengths. This image covers most of North America, all of Central America, and the extreme NW portion of South America. Because the GOES satellites maintain the same position relative to any point on the earth below, a view such as this provides the same coverage in any transmission as long as the satellite sub-point position is maintained. In this sample, a major cloud system blankets the southeastern US, while the Caribbean area is clear. This image was obtained in late February (see the header at the top of the image) and the lower ground temperatures of the northern states and Canada are clearly evident in the lighter tone of these areas.

ess is described in greater detail in Chapter 10.) The format now appears operational on all Meteor-3 series spacecraft.

The Russians also operate a higher-resolution satellite system that generates superb images at a rate of 240 LPM (4 lines per second). These pictures—usually transmitted from COSMOS satellites—are rarely copied in the western hemisphere because the satellites are often turned off by Russian ground controllers prior to loss of signal (LOS). European stations receive the pictures regularly when a COSMOS satellite is operational. US stations hear them on occasion when the satellites are left on to support Russian fleet maneuvers in the North Atlantic and North Pacific areas, and on other occasions where they remain on either accidentally, or as a result of changes in Russian ground-station operations.

GEOSTATIONARY SATELLITES

Geostationary satellites are in circular orbits over the equator at an altitude of approximately 22,000 miles. At this altitude, a single orbit of the earth takes 24 hours—precisely the time required for the earth to rotate once beneath the satellite. Thus, while the satellites are moving, the earth rotates below them at the same angular rate. In effect, the satellites remain over the same point on the equator and, from the ground,

they appear to remain at the same point in the sky. Once the proper antenna bearing for a specific satellite has been determined, the antenna can (except for some occasional adjustments) simply be locked in place for the operational lifetime of the satellite in question. The satellite signals are beamed back to earth at microwave frequencies, so a small parabolic dish antenna is usually used in conjunction with a suitable converter ahead of the station receiver.

The majority of the operational geostationary weather satellites have the primary mission of imaging their hemisphere using a multispectral scanning radiometer that provides very-high-resolution images at both IR and visible wavelengths every 30 minutes. The satellites spin on their axes at approximately 100 revolutions per minute (r/min), providing the horizontal scanning, while a motorized mirror with a period of approximately 20 minutes is used to provide the vertical scanning. This data is relayed back to earth in a very-high-density digital format that requires specialized equipment for reception and display. This image data is processed by high-speed computers on the ground that provide two different functions. First, the original data, obtained during the 30 ms of each 600-ms horizontal scan, is retransmitted in "stretched" form (at a lower bandwidth) during the 570 ms when the satellite sensors are scanning empty space. This

USA NOAA GOES-E 02/22/87 1800Z NE VS

Figure 1.7—An example of a GOES-E NE visible-light quadrant. This area of the earth was imaged at the same date and time as the IR example in Figure 1.6, but this sample shows the visible-light data. Visible-light quadrants such as this one can be recognized immediately by the fact that the view of space beyond the limb of the earth is black. Note, however, that there are many clouds in this view that were not obvious in the IR version in Figure 1.6. There is a major extension of the cloud system over the southeastern US extending across the Gulf to Yucatan. Only a few bits of this extension are evident in Figure 1.6. These clouds must therefore be quite low and warm, thus lacking contrast against the warmer Gulf waters. Similarly, clouds blanket Nicaragua and Panama, but are only slightly evident in the IR view for a similar reason. The same is also true for small cloud elements associated with Cuba and other islands in the Caribbean. A detailed comparison of visible and IR data can thus yield information on the vertical distribution of weather systems.

signal is in a digital format, and a number of amateur stations have succeeded in displaying this data (see Chapter 7).

The same ground computers also sector the data into individual quadrants, then relay these images in analog form back through the satellite as part of the Weather Facsimile (WEFAX) program. WEFAX transmissions use the same AM subcarrier as the polar-orbiter transmissions; the WEFAX signal format is described in detail in Chapter 4.

US WEFAX operations all originate from the GOES (*Geostationary Operational Environmental Satellite*) satellite series. The GOES network typically consists of three satellites. GOES-E, stationed at 75° W longitude, and GOES-W at 135° W longitude, are imaging satellites that transmit WEFAX pictures between primary image acquisitions. The third satellite, GOES-C, is situated at 107° W longitude and functions entirely as a WEFAX relay. GOES-C is usually an older satellite, retired from either the GOES-E or GOES-W position, and carries both GOES-E and GOES-W products. The positions given here are the nominal ones for a complete three-satellite network. Unexpected failure of one or more of the primary satellites can cause opera-

tional positions to be changed—sort of a space ballet—as functional satellites are nudged along the geostationary track to take up new positions to help compensate for the loss of a satellite.

The US primary GOES data is sectored in several ways for WEFAX transmission. The primary mode involves breaking the disc into four quadrants: northwest (NW), northeast (NE), southeast (SE) and southwest (SW). In addition, a tropical east (TE) and tropical west (TW), centered on the equator, are also available.

WEFAX transmissions derived from GOES data are of two types, representing visible-light and IR data. During the day, transmissions of visible-light quadrants are available. Grids outlining geographic and political boundaries, and longitude and latitude references (all added by the NOAA ground computers), make it easy to locate specific features (Figure 1.6).

Visible-light data formerly was not gridded in this fashion (Figure 1.7). With proper attention to the daily transmitting schedule, matching sets of visible or IR quadrants can be obtained, making it possible to reconstruct the entire earth disc.

If you observe visible-light and IR images from the

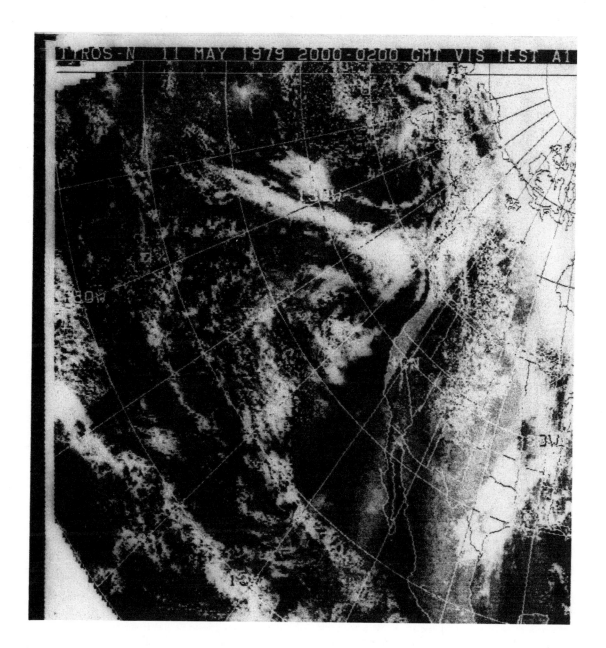

Figure 1.8—An example of a polar mosaic, prepared from TIROS visible-light data by the satellite-service ground computers and transmitted via GOES WEFAX. In this view, the north pole is at the upper right and out of the picture. This view covers western North America and a considerable area in the eastern Pacific, and was compiled from TIROS visible-light data. The image was printed on a version of the fax recorder illustrated in Chapter 4, using electrostatic recording paper.

same quadrant, obtained at the same time, they rarely look identical. First, the view of space beyond the limb of the earth appears black in the visible-light image, but white (cold) in the IR image. Quite a few cloud features that are apparent in the visible-light image may appear to be lacking in the comparable IR view. This is because IR-image contrast is a function of temperature differences. Cloud features close to the ocean surface are clearly evident in visible light, but because these low clouds are relatively close to land or sea surface temperature, they may be obscure or absent in the IR data. High, cold clouds appear very white in IR as opposed to lower (and warmer) clouds that appear as mid-range grays in IR images. These variations in tonal values allow you to reconstruct the vertical distribution of cloud systems in IR pictures, while a comparison of comparable visible and IR data provides the best overall view of weather systems.

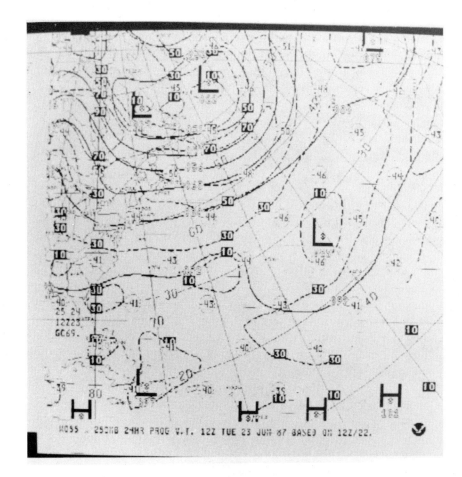

Figure 1.9—An example of a weather chart transmitted via GOES WEFAX. In this case, the chart is a polar projection (pole off to the upper left) covering the eastern US and Canada and most of the north Atlantic. This particular chart is a 24-hour VT (vertical temperature) prognosis. Chart ID data are contained in a footer along the bottom of the image.

In addition to the primary GOES image data, WEFAX is also used to transmit computer-generated Mercator or polar mosaics generated from visible and IR TIROS/NOAA polar-orbiter data (Figure 1.8). Because these products cover the entire world, it is possible to follow weather developments almost anywhere, regardless of your station location. WEFAX is also used to transmit weather charts (see Figure 1.9) as well as the WEFAX transmission schedule and polar-orbiter TBUS messages (see Chapter 8).

The European Space Research Organization operates a geostationary satellite, METEOSAT, (positioned at 0° longitude), which is quite similar to the US GOES operation in a number of respects. In the case of METEOSAT, the earth disc is broken down into a larger number of smaller quadrants. Although GOES WEFAX operations are confined to a single frequency (1691 MHz), METEOSAT operates on 1691 and 1694.5 MHz. The Japanese operate a similar satellite (GMS) over the eastern Pacific, but the wider RF bandwidth of these transmissions requires the use of larger antennas (see Chapter 3). Although the GOES WEFAX service was initially viewed as too complex for amateur use, advancements in technology and extensive amateur experimentation have made WEFAX ac-

cessible to almost any installation at relatively modest cost.

BASIC GROUND-STATION COMPONENTS

Most newcomers to weather-satellite activities start by assembling a polar-orbiter receiving installation. These satellites transmit on frequencies in the 136- to 138-MHz range, and require nothing more than a small VHF antenna (Chapter 2) and a simple VHF FM receiver of suitable design (Chapter 3). The picture-display system can take many forms, and several alternatives are discussed in Chapter 4, with a complete computer interface described in Chapters 5 and 6. Chapter 7 describes techniques for display of high-resolution digital imagery. In addition, you'll have to learn techniques for predicting satellite orbits to know when to expect passes from a given satellite and these are covered in Chapter 8. Integrating all of this into the overall operation of your receiving station is the subject of Chapter 9. Advanced techniques, such as image processing, are discussed in Chapter 10.

Upgrading to WEFAX requires the addition of a small microwave antenna (Chapter 2) and a downconverter (Chapter 3) to convert the microwave GOES signals to the 136- to 138-MHz VHF range. In most cases, a display system suitable for use with polar-orbiter pictures can be used directly or adapted for WEFAX display. The display system in Chapters 5 and 6 is completely compatible with WEFAX and all the present polar-orbiter modes. Antenna positioning calculations are covered in Chapter 8.

How much you spend on your satellite installation depends on which satellite signals you want to receive, how many features you desire, how much of the station you are willing to build, and whether or not you are adept at combing flea markets and surplus outlets for bargains. Completely functional polar-orbiter installations have been built for science-fair projects by Junior

High students for as little as $200. At the other end of the scale, you can easily invest several thousand dollars if you buy everything assembled and tested and insist on having the latest model of every possible piece of equipment.

Fortunately, this hobby is one in which some of your own skills and ingenuity can compensate greatly for a lack of ready cash. No matter what the cost of your station and no matter how simple or sophisticated it might be, there is endless potential for fun, fascination, and the daily excitement of watching the earth from space.

Chapter 2

Weather-Satellite Antenna Systems

INTRODUCTION

Your antenna system will consist of two primary elements. The first is the antenna, which is designed to intercept the small amounts of RF energy reaching your location from the distant satellite. The second system component is the transmission line, a cable designed to carry these faint signals from the antenna back to your receiver, hopefully while minimizing signal losses in the transfer. Most of this chapter is devoted to the antenna portion of the system, but I'll have a general discussion of transmission lines and the all-important cable connectors at the end of this chapter.

Many weather-fax enthusiasts will find the installation of an antenna (or antennas) to be the single greatest problem they have to solve in getting a weather-satellite station operational. Increasing numbers of condominiums, urban apartments, and suburban neighborhoods with restrictive real-estate covenants may make it appear that there is no option for an effective weather-satellite antenna system. If this is your situation, there is hope. A new section, **OPTIONS**, has been added at the end of this chapter. In it you should find one or more solutions that will work for you no matter how restrictive your living situation might be.

ANTENNA BASICS

Gain

Gain is a measure of how much a given antenna increases the level of a signal relative to some reference standard—usually a simple dipole. The gain units are decibels (dB) and, since the scale is logarithmic, a gain of 3 dB equates to a doubling of the signal strength. Gain is highly desirable since it helps to overcome cable losses and makes the RF preamplifier in the receiver less critical. Unfortunately, to get an increase in gain you have to trade off another antenna parameter—the beamwidth.

Beamwidth

Beamwidth is a measure of the width of the antenna pattern. Generally, an antenna with low gain has quite

a wide pattern, receiving signals well from a number of different directions. If we want to increase gain, we must do it at the expense of decreasing the beamwidth. A high-gain antenna must be pointed rather accurately, otherwise the received signal will be quite a bit weaker than would have been the case with a low-gain/wider-beamwidth antenna. A high-gain VHF antenna for one of the polar-orbiter satellites will easily deliver a very strong signal to the receiver, but to do so, it must be accurately tracked during a pass so that it remains pointing at the satellite at all times. In contrast, an omnidirectional antenna designed to accept signals from all directions may deliver a comparatively weak signal, but does not require tracking the satellite!

Polarization

The orientation of radio waves in space is a function of the orientation of the elements of the transmitting antenna. In our daily lives, we encounter two principal modes of polarization—*horizontal* and *vertical*. FM broadcast and TV transmissions are typically horizontally polarized, and an antenna designed for such signals has elements oriented horizontally. In contrast, police, other public service, and mobile (cellular) phone transmissions use vertical polarization because a simple vertical whip antenna is the easiest sort of omnidirectional antenna to mount on a vehicle. If a horizontally polarized signal is received on a vertical antenna, or vice versa, we refer to this as *cross polarization*. Cross polarization is highly undesirable since signal losses can exceed 20 dB—more than enough to render an otherwise strong signal completely unreadable.

Both vertical and horizontal polarization are examples of *linear* polarization and present real problems with space communications—particularly in the case of polar-orbiting satellites. Such satellites are in constant motion with respect to the ground station. Consequently, a linearly polarized signal from a satellite appears to be constantly changing polarization with respect to the ground station. The result can be strong signals at one time and very weak or unreadable signals just minutes later. Because linearly polarized antennas are simple and easily mounted on spacecraft, early satellites used them; the Meteor satellites still do. It was therefore necessary to solve the variable polarization problem at the ground-station end: The answer is to use *circularly* polarized antennas.

A circularly polarized wave rotates as it propagates through space, and antennas can be designed for either right-hand circular (RHC) or left-hand circular (LHC) polarization. An RHC- or LHC-polarized antenna shows a maximum loss of only 3 dB when receiving a linearly polarized signal, so circularly polarized antennas are almost universally used at polar-orbiting-satellite ground stations. Either an RHC- or LHC-polarized antenna would do fine for Meteor reception, but if we are going to also receive the TIROS/NOAA satellites, we must use an RHC-polarized antenna because the TIROS satellites generate an RHC-polarized signal—not a linearly polarized one. If we receive the TIROS signal on an RHC-polarized antenna, we experience no polarization loss. Should we try to receive the RHC-polarized signal using an LHC-polarized antenna, we would experience cross-polarization and have a constant 20-dB polarization loss to contend with. Thus, our best bet is to use an RHC-polarized antenna where we'll have no TIROS polarization losses and a maximum of 3 dB of Meteor signal loss.

A linearly polarized antenna can be used with the RHC-polarized transmissions from NOAA satellites with a maximum polarization loss of only 3 dB. Although such an antenna is much simpler than an equivalent one designed for circular polarization, it is unsuited for use with Meteor satellites. Some additional options with regard to antenna polarization are discussed at the end of the *Gain Antenna* section.

Geostationary satellites such as GOES do not move in relation to the ground, so linear antenna polarization can be used at both ends of the circuit.

Unfortunately, the two major frequency ranges we must use—VHF for polar orbiters and S-band for geostationary satellites—require that we use quite different antennas for each service. The VHF antennas are much like conventional TV antennas in terms of their design, but S-band antennas are almost always of the parabolic dish type, similar to those used for satellite-TV reception. There are other types of S-band antennas, but dishes are simpler to work with unless you have access to some pretty sophisticated test equipment.

VHF ANTENNA SYSTEMS

One of the most common forms of VHF antenna is the *Yagi*, named in honor of the Japanese scientist who first elucidated the principles of combining a basic dipole—a so-called *driven element*—with a number of *parasitic elements*. Although the driven element is connected to the transmission line, the parasitic elements are coupled to the driven element by resonance, not by any physical connection. Parasitic elements are of two general types. *Directors*, which are electrically shorter than the driven element, are placed in front of the driven element, facing the source of RF energy in the case of a receiving antenna. Most Yagis also employ a single *reflector* that is electrically longer than the driven element and is placed behind it. The driven element, in combination with one or more parasitic elements, makes up a beam antenna, of which your common TV antenna is a prime example. A simple beam usually starts with a driven element and a single reflector. As parasitic elements are added (usually in the form of an increasing number of directors) gain increases, beamwidth decreases, and generally, the frequency range over which the antenna will operate (bandwidth) becomes narrower, although this is far less serious for receiving applications than it is for transmitting.

To achieve circular polarization, a beam antenna can be built using two identical sets of elements, each mounted at right angles to the other. The two driven elements must be properly phased with a length of transmission line between them and, depending upon how the connections are made, either RHC or LHC polarization can be achieved. The four-element Yagi described in the *Gain Antenna* section is one easy approach to home construction. There are other types of directional VHF antennas that are documented in a variety of ARRL publications.

Any kind of directional antenna requires tracking to keep the antenna pointed at the satellite as it moves across the sky. This requires two antenna rotators—one to control the direction (or azimuth) in which the antenna is pointed and the second to control the elevation or the angle at which the antenna is pointed relative to the horizon. This hardware requires that you use a manual plotting board or a computer to determine the precise times for various azimuth and elevation settings to keep the antenna pointed at the satellite during a pass.

For most of us, the use of a directional antenna means we are confined to weekend and evening passes when we can operate the tracking system. There are approaches that can be used to automate the tracking process to permit unattended recording of satellite passes, but they'll increase the overall cost of your station. The popularity of Amateur Radio satellite communications has resulted in a number of manufacturers producing sophisticated rotators and control systems for moving antennas in both azimuth and elevation. Some of the systems can even be controlled by a remote computer. Such a rotator control system costs about $500, plus the cost of the computer system—if you want to take advantage of all of the flexibility such a system can provide.

For many operators, the antenna needs for VHF operation can be met with an omnidirectional antenna system. These antennas are simple and inexpensive to construct and have no need for azimuth or elevation rotators. Such an antenna will deliver noise-free pictures for the best passes of a given satellite (which occur twice each day), including the "worst-case" situation where a set of passes straddles the ground-station location. The major advantage of such an antenna system is that no tracking is required and unattended operation can be as simple as connecting a timer to the station tape recorder.

Unless you have the need to regularly access passes at extreme range, there is really no need for a beam antenna with all the additional complexities that are involved. Most of the TIROS/NOAA and Meteor images in this book were obtained using the omnidirectional antenna to be described in the next section, and graphically illustrate its effectiveness. In the case of overhead passes, I can hold the signal at full quieting from Ellesmere Island, north of Hudson Bay, to just north of Yucatan. In the case of worst-case straddling passes, the coverage is reduced from central Hudson Bay to just south of Florida.

AN OMNIDIRECTIONAL VHF ANTENNA

When I published an article, shortly after the appearance of the first edition of the *Handbook*, which I titled "An Omnidirectional Circularly-Polarized Antenna for Weather-Satellite Reception," a practical editor at *73 Magazine* shortened the title to the "Satellite Zapper." Despite my best efforts to return to the pompous original, it remains to this day as the "Zapper." The Zapper is simple in concept in that it is nothing more than a short beam (two driven elements and two reflectors) with such a wide beamwidth that, when pointed straight up, it functions as an omnidirectional antenna system that does not require any tracking for passes within your "best-pass" window. With a hot receiver and relatively short transmission line, the Zapper performs quite well, although it is at its best when a low-noise preamp—either a JFET (*junction field-effect transistor*)—or, better still, a GaAsFET (*gallium arsenide field-effect transistor*) is mounted at the antenna. With a preamp in place, the length of transmission line between the antenna and receiver is basically irrelevant.

The Zapper is an ideal first project because of its simplicity of construction and ease of mounting. If you later use a gain antenna to maximize your coverage at the limits of reception, you'll still use the Zapper regularly for general monitoring, spotting new Soviet satellites by scanning various frequencies, etc.

A few words might be in order for those of you who are familiar with VHF-antenna design. If you are a purist, you should be prepared to be horrified by some aspects of the Zapper's design. You may be tempted to "clean up" the design by following accepted rules for such things as matching, and so on. Please keep in mind that we are not dealing with transmitting antennas; instead, we're looking for the best possible reception with the simplest possible approach to getting the job done. The earliest pre-Zapper started out following all the rules—and it didn't work particularly well. With each revision, the design became simpler and performance increased. The newest version shown here is the simplest and best yet. You can clean it up if you want to, but be advised that it may not work as well as this one. The evolution from orthodox to unorthodox has been quite purposeful, and you should keep that in mind before embarking on major "improvements."

The Zapper requires a vertical mast, two reflectors, and two driven elements. The driven elements mount at right angles to each other at the top of the mast, offset vertically by a distance of two inches. The reflectors mount approximately ¼ wavelength below the driven elements, with each reflector parallel to a driven element. The overall antenna is quite compact and unobtrusive. This is an important factor in many areas where restrictive deeds and real-estate covenants can severely restrict the kind of antennas that can be erected.

Materials

The original Zapper, which appeared in the second and third editions of the *Handbook*, was fabricated from various sizes of aluminum tubing. Such tubing can be hard to obtain and is relatively expensive from most sources. Assembling the antenna required careful drilling of holes that never seemed to line up properly, and the hardware always seemed to corrode, no matter how carefully the antenna was weatherproofed. For this edition, I redesigned the Zapper (let's call it Zapper II) so that it requires no aluminum, no drilling, no assembly hardware, and it won't corrode. It works as well as the original, costs less, and looks quite a bit better hanging up in the breeze.

The secret to this new version of the antenna is the use of standard ½-inch CPVC plumbing pipe and fittings that you can obtain from almost any hardware or discount store. (From here on, I'll refer to CPVC simply as PVC.) The antenna-element housings and their supporting framework are constructed entirely of PVC pipe and fittings, providing both the rigidity required and complete weatherproofing of the actual antenna elements. To build the antenna you'll need the following materials:

2   10-ft (3 m) lengths of ½-inch (1.27-cm) PVC pipe
8   PVC **T** fittings for ½-inch pipe
11 ½-inch PVC end caps

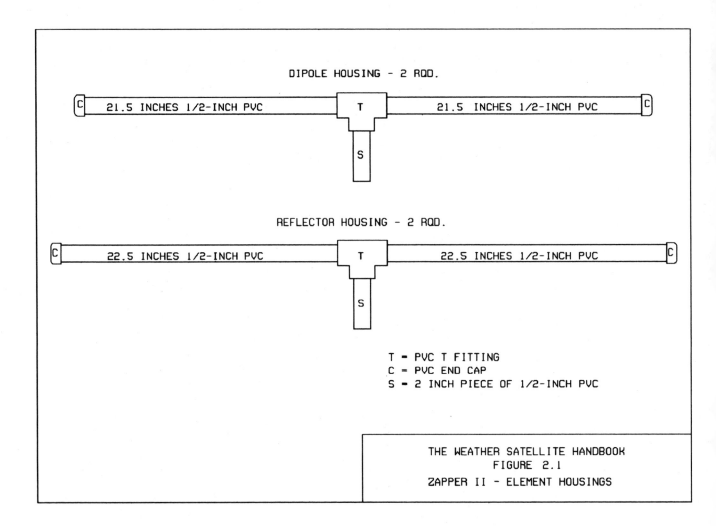

DIPOLE HOUSING - 2 RQD.

REFLECTOR HOUSING - 2 RQD.

T = PVC T FITTING
C = PVC END CAP
S = 2 INCH PIECE OF 1/2-INCH PVC

THE WEATHER SATELLITE HANDBOOK
FIGURE 2.1
ZAPPER II - ELEMENT HOUSINGS

1 bottle of PVC pipe cement compatible with your pipe
1 5-to 6-foot (ca. 2-m) length of aluminum TV mast
4 stainless-steel hose clamps

In addition, you'll require the following antenna items from your local electronics outlet or mail-order outlet:

1 20-ft length of RG-58 coaxial cable (Belden 8219 or equivalent), plus enough additional cable for the run from the antenna to your station location
1 8-ft length of 300-ohm TV twinlead
2 double-female BNC adapters
5 BNC plugs for RG-58 cable and an additional connector to match your receiver antenna-input jack
1 BNC **T** adapter (female arms, male common)

Use a hacksaw to cut the following lengths of ½-inch pipe:

3 1 inch (2.6 cm)
4 2 inch (5.2 cm)
4 21.5 inch (54.5 cm)
4 22.5 inch (57 cm)
1 8 inch (20.32)

Use a file to deburr the cut ends of the tubing and put them aside with the PVC fittings.

Reflectors

We'll start antenna assembly with the reflector elements because they're the simplest to fabricate. In the steps that follow, temporarily assemble the indicated pieces and check to make sure you have them properly aligned. When you're ready to make the assembly permanent, coat one end of the indicated piece of PVC tubing with the tubing cement and insert the piece firmly into the indicated fitting.

Using the reflector housing diagram in Figure 2.1, cement two 22½-inch pipe sections into the side arms of one of your PVC **T** fittings. Cement a 2-inch length of pipe into the lower position of the fitting, then cement a PVC cap onto one end of the housing, leaving the other end open. Repeat this entire assem-

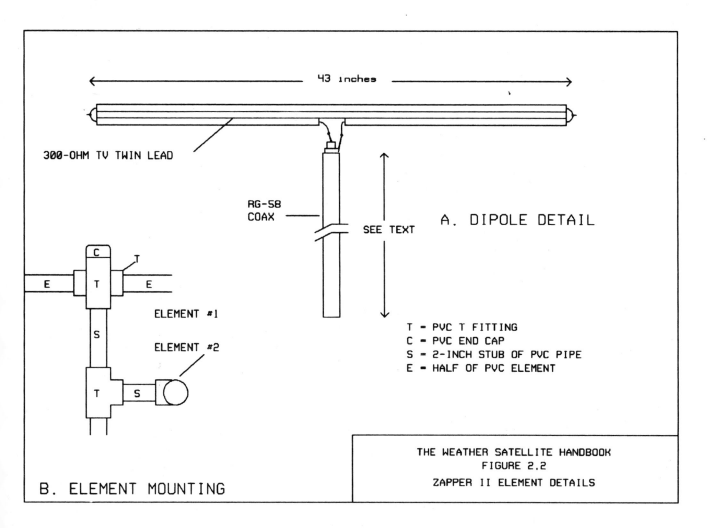

43 inches

300-OHM TV TWIN LEAD

RG-58 COAX

SEE TEXT

A. DIPOLE DETAIL

T = PVC T FITTING
C = PVC END CAP
S = 2-INCH STUB OF PVC PIPE
E = HALF OF PVC ELEMENT

C
T
E   T   E
S
ELEMENT #1
ELEMENT #2
T   S

B. ELEMENT MOUNTING

THE WEATHER SATELLITE HANDBOOK
FIGURE 2.2
ZAPPER II ELEMENT DETAILS

Figure 2.2—Zapper II element details. At A, one conductor of a piece of twinlead has been cut and soldered to a length of coaxial cable. At each end of the twinlead, the two conductors are twisted together and soldered (see text). At B, the method of mounting two elements at right angles to each other. In this view, you're looking at the back of the top **T** (element #1), with a PVC pipe cap at the top. A two-inch stub connects the lower end of the top **T** to a second **T**. The second **T** is attached to a third **T** by another two-inch stub. The third **T** (part of element #2) is viewed end-on from the capped end of the element.

bly sequence with a second set of pieces and set both reflector housings aside for at least one hour to let the cement cure.

Cut two 45-inch lengths of RG-58 coaxial cable. When the reflector housings have set, insert one piece of the cable down the length of each housing, cement a PVC cap to the open end, and lay the assembly aside.

The following assembly steps refer to Figure 2.2B. Insert a 1-inch length of tubing into one end of the cross-arm of a **T** fitting and cement a cap to the end of this stub. Insert a 2-inch piece of pipe into the opposite end of the fitting. Insert the pipe stub from one of the reflector housings into the side arm of this fitting so that the element is at right angles to the long axis of the fitting, as shown in the upper part of Figure 2.2B.

Your orientation should be accomplished quickly because the cement sets rapidly.

Add a second **T** fitting to the free end of the 2-inch stub, oriented at right angle to the upper fitting. Insert a 1-inch length of pipe into the bottom of this lower fitting and cap it. Now insert the 2-inch stub from your second reflector housing into the side arm of the lower **T** fitting. You should end up with your two reflector housings at right angles to each other and separated by about 2 inches. Lay aside the complete reflector assembly, but keep it near at hand so you can refer to it as you assemble the driven-element housings.

Driven Elements

Construction of the two driven elements begins with the assembly of two dipoles, illustrated in Figure

Weather-Satellite Antenna Systems   2-5

2.2A. Cut two 4-foot (1.2-m) lengths of RG-58 cable. Carefully cut through the plastic jacket at a point about 2/3 inch (2 cm) from one end of each piece. Score the jacket lengthwise from this point and peel the jacket away. Unravel the shield braid, then twist it together into a single, short stub. Remove the insulation from the upper half of the exposed center conductor. Use a soldering iron to tin both the center conductor and the tip of the braid stub.

Cut two 44-inch (111.8-cm) lengths of the 300-ohm TV twinlead. For each piece, cut enough of the webbing from between the conductors at each end so that you can strip the two conductors, twist them together, and solder the connection. Mark the middle of each piece and cut through one of the conductors at that point. Strip 1/2 to 1 inch (1 to 2 cm) of insulation from the wires on each side of the cut. For each piece, solder the center conductor of one of the pieces of coax to one wire and the braid to the other. You should now have two dipole assemblies like the one illustrated in Figure 2.2A.

Dipole Housing

We'll now build the PVC housing around each of the dipoles you have just constructed. Feed the two ends of the dipole through the base of a **T** fitting, routing one side of the dipole out one side arm opening and the other out through the remaining opening. Work the individual dipole legs outward from the fitting until the solder connections to the coaxial cable are located up inside the **T** fitting. Slide one 21 1/2-inch piece of pipe over one leg of the dipole and cement it to the **T** fitting, repeating the operation with the other dipole leg and a second 21 1/2-inch piece of tubing. Cap the open ends of the tubing. Now take a 2-inch length of pipe, slide it up the coax and cement it to the remaining opening in the fitting. You should now have a pair of dipole assemblies, enclosed in the PVC pipe with the coaxial cable coming out the 2-inch pipe stub on each housing.

All that remains now is to use the remaining caps and **T** fittings to make a complete driven element assembly (two elements) exactly as you did with the reflectors. The orientation of all the pieces should be just the same as the reflector assembly. Because the lengths of coax have to be threaded through the **T** fittings, I suggest the following assembly sequence:

1) Thread the free end of the coax from one dipole through the side hole of a **T** fitting (assuming the final vertical orientation of the **T** fitting) and out the lower hole.

2) Cement the 2-inch dipole stub to the side hole, orienting the fitting at right angles to the dipole assembly.

3) Repeat the preceding two steps with the second dipole assembly and the remaining **T** fitting.

4) For the uppermost dipole, insert a 1-inch pipe stub in the upper hole and cap it.

5) Slide a 2-inch piece of tubing up the coax of the upper dipole and cement it to the lower hole of the **T** fitting where the coax exits.

6) Feed the upper coax cable through the upper hole of the lower **T** fitting and out the bottom hole with the coax from the lower dipole, cementing the pipe from the upper fitting to the upper hole of the lower fitting. Be sure the two dipoles are at right angles to each other before the cement has a chance to set.

7) The lower **T** fitting should now have two cables exiting from the lower hole. Slide the 8-inch pipe piece up both cables and cement it to the lower **T** fitting.

After the dipole assembly has set, install BNC connectors on the free ends of the two lengths of coax. One cable (the one from the lower dipole) will appear longer than the other. Do NOT trim them to equal length when installing the connectors!

Cut a 13-inch length of coaxial cable and install BNC connectors on each end. This piece of cable is our phasing line.

Final Mounting

Stainless-steel hose clamps are used to mount the dipole and reflector assemblies on the length of aluminum pipe or TV mast. If circumstances permit, it's usually easier to mount the assemblies with the mast in place. Alternatively, you can mount them on the mast and then install the mast section. The dipole assembly should go at the top of the mast, with the ends of the upper dipole facing E-W. The reflector assembly should be mounted so that the upper reflector is parallel with the upper dipole and 17 inches (44 cm) below it. Your mounted antenna should look similar to that shown in Figure 2.3—but with the reflector and driven elements properly oriented!

For the best possible results, the antenna should be mounted as high as possible with a minimum of obstructions within 5 degrees of the horizon. If a preamplifier is mounted at the antenna, the length of transmission line between the antenna and receiver is irrelevant, allowing you to choose the best possible site for the antenna without worrying about transmission-line losses.

Cables

The longer cable from the dipole assembly should be connected to one arm of the BNC **T** connector. Connect the phasing line to the other arm and use a double-female BNC connector to connect the free end of the phasing line with the shorter cable from the dipole assembly.

For the best possible performance, the preamplifier should be mounted at the antenna with a housing to

Figure 2.3—A photograph of the prototype of the Zapper II antenna. The antenna is quite unobtrusive and can be installed using a small tripod or other minimal support. You should take care where the assembly notes indicate that the reflector and driven element assemblies have the same orientation. With this prototype I didn't, and ended up with the lower reflector on the other side of the mast from the lower driven element. Although this should skew my pattern a bit to the east in one plane, there doesn't seem to be any obvious effect.

keep rain or snow off the unit. A suitable housing can be made from a plastic freezer container, secured to the mast with hose clamps. The top of the container should face down and the cover should be equipped with holes to let the cables pass up from below. These holes also vent the case so that condensation does not build up inside the enclosure. The male connector of the BNC **T** fitting can connect directly to the preamplifier input, or you can use a short length of interconnecting cable. The transmission line to the station receiver can connect directly to the preamplifier output. If your preamplifier is powered through the transmission line, your installation is now complete. Otherwise, you'll need to run a wire with the transmission line to carry 12 V dc to the preamplifier. A ground wire is not required because the coaxial cable shield provides the ground return.

Although the antenna performs best with a mast-mounted preamplifier, a direct run of cable (using another double-female BNC adapter) can be connected to the BNC **T** connector. For best results, the receiver should have a low-noise front end, and no more than 50 feet (15 meters) of transmission line should be used—preferably less.

Finishing Touches

Prior to taking care of the final details, run a reception check on the antenna system. Using a vintage Vanguard receiver and a Hamtronics GaAsFET preamplifier, I can expect to hear a satellite on the horizon with full quieting by the time the satellite has reached an elevation of 3 to 4 degrees, assuring a minimum of 12 to 14 minutes of coverage for a worst-case situation with a pair of straddling passes. With the exception of a few examples of West Coast passes, all the polarorbiter images in this book were obtained with the omnidirectional antenna system.

As a final check on polarization, observe a number of NOAA passes. If there are deep nulls in the pattern, as evidenced by noise at moderate to high satellite elevation, switch the phasing line position from the short to the long cable.

Once you know the antenna is working, use plastic electrical tape to cover all exposed connectors. Once they have been taped, cover them with the putty-like sealer available at most electronics shops that carry TV or CB antenna supplies, and tape them again. The sealer can also be used to seal the base of the dipole assembly where the cables exit. Electrical tape can then be used to secure the cables neatly to the mast. A coat or two of paint will protect the tubing from the effects of UV radiation and assure an almost unlimited service life for your antenna assembly.

One area where you can experiment concerns the spacing between the driven element and reflector assemblies. The 17 inches (57 cm) specified represents approximately 0.2 wavelength at 137 MHz, and is a good starting point. Increasing this spacing changes the pattern slightly in ways that might prove useful to you. At a maximum spacing of ⅜ wavelength (32.25 inches [82 cm]), the gain directly overhead is decreased slightly, but gain improves closer to the horizon. You should not exceed this value of ⅜ wavelength because beyond that point, the pattern breaks down from a single broad lobe to a series of minor lobes that will lead to very erratic reception.

Commercial Options

Omnidirectional antennas similar to the Zapper have now become quite fashionable and a number of vendors offer designs that represent an alternative to building your own. Most feature integrated weatherproof preamps as options or as standard equipment.

Figure 2.4—The WEFAXTENNA from Vanguard Labs, an example of a moderately priced commercial omnidirectional antenna. This antenna uses a pair of phased dipoles over a ground plane consisting of 8 radial elements. The mast is a piece of thick-walled PVC tubing that provides weather protection for the dipole phasing lines as well as the preamplifier which is powered through the transmission line. The antenna is available with or without the preamplifier.

Figure 2.4 illustrates one commercial variant similar to the Zapper concept. More exotic antenna options, such as the quadrifilar helix from Satellite Data Systems, provide very good results when used as an omnidirectional ground-station antenna.

VHF GAIN-ANTENNA SYSTEMS

Although the Zapper gives a good account of itself for the best passes of the day, tracking a satellite out to the eastern or western horizon requires a gain antenna system. Figure 2.5 (which also appears on the cover of this edition) is a mosaic prepared from two passes from NOAA-11 on 2 Sep 1989. The eastern pass, extending from northern Yucatan to northern Hudson Bay, was a near-overhead pass slightly to the east and was logged with the Zapper antenna. The western pass is another story, for the maximum satellite elevation was about

13 degrees. Although the Zapper would have provided perhaps 6 minutes of coverage for that pass, the use of a small 5-element beam permitted me to get over 12 minutes of coverage, resulting in a spectacular mosaic of most of North America.

When receiving satellite images first became popular, the VHF antenna arrays tended to be fairly complex, involving at least 10 vertical and 10 horizontal elements phased for circular polarization. The problem in those days was to get enough gain to overcome the poor noise figures of the available RF preamplifiers. Such antennas had very sharp patterns that required accurate tracking of the satellite to take advantage of the gain provided by the antenna system. Today, with the universal use of GaAsFET preamplifiers, a far more modest antenna system yields horizon-to-horizon coverage with a far broader pattern, leading to a considerable relaxation of the required tracking accuracy.

A Basic Four-Element Beam

A four-element beam provides a useful basic building block for a variety of final antenna options. If you want to build such an antenna, it is possible to use the PVC pipe approach employed for the Zapper, or you can work up the same basic antenna using aluminum tubing and more conventional antenna construction. Let's look at the PVC version first.

The essential dimensions of the beam are shown in Figure 2.6A. The beam is constructed using PVC pipe and fittings, just like the Zapper. You'll need two 10-foot (3-m) pieces of ½-inch pipe, 8 T fittings, and 9 end caps. Cut your tubing into the following lengths: four 20½ inch (52 cm), two 21½ inch (54.6 cm), two 22½ inch (57.2 cm), two 16 inch (41 cm), one 22 inch (56 cm), and five 1 inch (2.5 cm).

Follow the Zapper instructions for the construction of one reflector and one driven element. The only deviation involves the use of 1-inch pipe stubs for these elements instead of the 2-inch units specified for the Zapper. This permits the T fittings of the elements to be mounted flush with the T fittings of the boom when we get to final assembly.

You'll also need to construct two directors. These are assembled just like the reflector except that each uses a 41-inch (104-cm) length of coax, two 20½-inch (52-cm) pipe pieces, a T fitting, a 1-inch pipe stub, and two end caps.

The boom consists of four T fittings interconnected by PVC pipe sections. The reflector and driven element T fittings are interconnected by a 17-inch (41-cm) pipe section, the driven element and first director fittings are linked with another 17-inch (41-cm) pipe, and a 22-inch (56-cm) section links the fittings for the first and second directors. Insert a 1-inch (2.5-cm) stub at the front (second director end) of the boom and

Figure 2.5—A two-pass NOAA-11 mosaic covering almost all of North America. The eastern pass was essentially overhead at my location and was logged with the Zapper II antenna. The western pass had a maximum elevation of 13° and was obtained by manually tracking the satellite with a small, 5-element Yagi antenna. Combining two passes like this is a bit trickier than simply linking the northern and southern pieces of a single pass. Note, for example, that the passes do not simply lie neatly side by side. In fact, they converge toward the poles, as your tracking exercises in Chapter 8 will indicate. Although you can expect to achieve a near-perfect match with geographic features, the match is likely to be a problem with clouds, especially toward the south. This is because the two passes are separated by about 102 minutes—almost two hours—during which the cloud systems will move! Differential lighting can also be a problem. Note that the region north of Lake Superior is quite dark in the eastern pass because of low light levels. The same latitude in the later western pass has more favorable lighting conditions. Such variations could be compensated for, to some extent, in the darkroom, but I did not make that effort in this case.

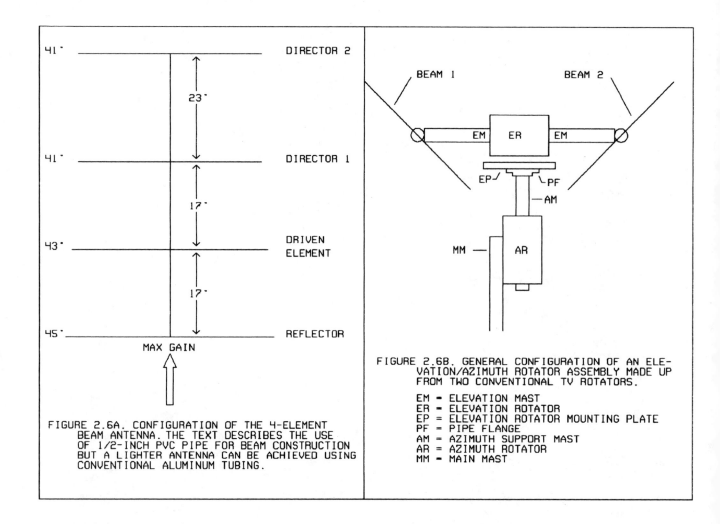

FIGURE 2.6A. CONFIGURATION OF THE 4-ELEMENT BEAM ANTENNA. THE TEXT DESCRIBES THE USE OF 1/2-INCH PVC PIPE FOR BEAM CONSTRUCTION BUT A LIGHTER ANTENNA CAN BE ACHIEVED USING CONVENTIONAL ALUMINUM TUBING.

FIGURE 2.6B. GENERAL CONFIGURATION OF AN ELE-VATION/AZIMUTH ROTATOR ASSEMBLY MADE UP FROM TWO CONVENTIONAL TV ROTATORS.

EM = ELEVATION MAST
ER = ELEVATION ROTATOR
EP = ELEVATION ROTATOR MOUNTING PLATE
PF = PIPE FLANGE
AM = AZIMUTH SUPPORT MAST
AR = AZIMUTH ROTATOR
MM = MAIN MAST

cap it. Mount the two directors and the reflector at right angles to the boom, then mount the driven element, threading the coax out the hole at the rear of the boom. Adding a connector to the cable and sealing the exit hole for the cable completes the beam—except for mounting hardware.

There are two basic disadvantages to the PVC version of the antenna. First, the boom has to be stiffened (see *Installation Options*) with a piece of aluminum angle stock. Second, the beam is relatively heavy compared to an aluminum version of the same antenna. These two disadvantages are offset by the ready availability of the materials compared to aluminum tubing.

The basic dimensions in Figure 2.6A can be used as a guide to duplicating the beam using aluminum tubing if desired. *The ARRL Antenna Book* shows several approaches to VHF Yagi construction and all the 2-meter antennas can be modified according to the dimensions provided.

Although several companies do market Yagi antennas cut for 137 MHz, most are somewhat expensive because they are produced in relatively small quanti-

ties. The same manufacturers typically market small Yagi designs for the amateur 2-meter (144 to 148-MHz) band. These antennas usually cost less because they are produced in much larger numbers and the manufacturers are competing for sales. What is generally not realized is that these 2-meter antennas will work almost as well (receiving, *not* transmitting!) as an antenna cut for the 136- to 138-MHz satellite band. A small 2-meter antenna will show good gain and directivity when used for weather-satellite reception, and the antennas are readily available from Amateur Radio supply houses throughout the world, as well as directly from the manufacturers. Typical models that do a good job in this service include:

Vendor	Model	Configuration	List Price
Cushcraft	A147-4	4-element linear	$50
	A144-10T	5+5 circular	$85
Hy-Gain	205S-1 25BS	5-element linear	$45
KLM	2M-4X/144-148	4-element linear	$62

If you are really interested in saving money, check with local radio amateurs. It is quite common for them to have one or more old 2-meter beams stashed in the garage or other storage area. One of these old antennas, cleaned off with steel wool and with new element-mounting hardware, does an excellent job receiving weather-satellite signals—and you may be able to get the antenna for free!

If you ask around, you'll probably hear that 2-meter beams cannot possibly work in the 136- to 138-MHz range. The fact is that they do! The western half of the mosaic in Figure 2.5 was received using a *very* old 5-element 2-meter beam. As a new radio amateur (WV2EMC/WA2EMC) back in 1958, in northern New Jersey (living, believe it or not, in a log cabin on one of those little lakes that you can see in Figure 1.1), I was given the beam by W2KHQ so I could log onto the 2-meter Civil Defense net. The antenna is certainly over 35 years old, yet was still residing in the back of the garage after all those years and a great many intervening moves. Disassembled, cleaned, and with replacement hardware, the antenna works just fine!

Installation Options

The 4-element beam, as described, is a linearly polarized antenna with a gain of approximately 9 dB over a simple dipole. When used in any orientation to receive TIROS/NOAA signals, the actual gain is about 6 dB due to the 3 dB loss experienced when receiving the RHC-polarized satellite signal. This gain is entirely adequate. The various linear 2-meter antennas are roughly comparable. You can install such an antenna on a commercial azimuth-elevation (az-el) rotator system (such as the Yaesu G-5400B), or you can construct your own az-el rotator system using two standard TV antenna rotators.

The single antenna won't suffice for Meteor reception because these satellites use linear antennas. The easiest approach to achieving circular polarization is to build a second 4-element Yagi and orient it so that its elements are at right angle to the first Yagi. The phasing line and BNC **T** connector can then be used to interconnect the two antennas following the description for cabling the Zapper.

Figure 2.6B shows the general configuration for an az-el rotator system using two TV rotators. The azimuth rotator is mounted to your main support mast in the usual TV fashion, and its control box is used to control the antenna azimuth or compass direction, just as it would for normal TV service. The critical difference is that instead of turning the antenna directly using a short section of mast in the azimuth rotator, the rotator turns a short pipe stub with a pipe flange secured to a flat, metal plate. The second rotator, which is used for elevation control, is mounted, on its side, to this plate. A short section of mast from this second rotator mounts to the antenna, allowing that rotator control box to move the antenna up and down.

Many of the details of constructing your rotator system will depend on the TV rotators available to you. Most rotator housings are weatherproofed, but since they are not designed to be mounted on their sides, the sealing may not be completely effective for the elevation rotator. It may be worthwhile to fabricate a splash guard of sheet metal or plastic to keep the worst of the rain off the elevation-rotator housing. A rotator in which the movable mast extends completely through the housing is the best overall option because it permits the use of a counterweight to balance a single antenna, or lets you mount two antennas for circular polarization. A rotator without this feature can still be used for elevation control of a single antenna if the horizontal mast extension to the antenna is not more than about 2 feet (70 cm) long.

Mounting the PVC beam antenna requires that you stiffen the boom. This can be accomplished with a piece of ½-inch aluminum angle stock (cut to the length of the boom), using stainless-steel hose clamps to secure the boom along the length of the angle stock.

Mounting the antenna is done using an aluminum plate about 4 inches (10 cm) square. The plate should be fairly stiff so a thickness of at least ⅛ inch (3 mm) is indicated. Use a pair of **U** bolts to secure the boom to the plate at the balance point for the antenna—somewhere between the driven element and first director. Another pair of **U** bolts is then used to secure the plate to the mounting mast so that the boom is at right angles to the mast. If a single antenna is used, the elements should be oriented vertically to minimize the interaction with the mast and rotator. If you are mounting two antennas for circular polarization (necessary only for Meteor reception), mount each at 45 degrees to the horizontal mast and 90 degrees to each other, as shown in Figure 2.6B. If you are using one of the 2-meter antennas, they should already have mounting hardware installed, making it an easy job to mount one or two of them.

If your situation does not permit you to mount such an antenna system, or you are only interested in casual operation and don't wish to expend the time and money for a permanent installation, check out some of the suggestions in the **OPTIONS** section at the end of the chapter.

S-BAND GOES/METEOSAT ANTENNAS

The importance of adequate gain in the S-band receiving system will be highlighted in Chapter 3. In terms of antenna gain at these frequencies, the major determinant is the diameter of the dish. The job of the dish is to reflect the incoming RF energy to a focal point where it can be picked up by a feed horn or other device to transfer the RF energy to a transmission line. The bigger the dish, the greater will be the intercepted RF energy and hence, the greater the gain! Usable GOES/METEOSAT dishes come in quite a range of sizes:

Diameter			
Feet	Meters	Source	Gain (dBi)
2	0.6	Saucer sled	18
4	1.2	Commercial/surplus	24
6	1.8	TVRO	27.5
10	3.0	TVRO	32

At the small end of the size range (2-3 feet), we can have a workable and inexpensive antenna, but there are two major drawbacks. First, the minimal gain puts great demands on the RF preamplifier—and there is no margin for any appreciable line losses between the antenna and the preamp input. Secondly, the small antennas have such a wide pattern that at certain elevation and azimuth angles you may experience interference from adjacent satellites. I have a 2-foot dish that I use for demonstrations and portable work, but if I train this small dish on GOES W, I also get interference from GOES C! Still another problem with very small antennas is that if a problem develops with a satellite that results in a drop in signal level, you may no longer have enough of a signal margin for noise-free reception.

A 4-foot (1.2-m) dish (Figure 2.7) is about optimum for combining small size with adequate gain and a narrower beamwidth. Larger dishes will provide a greater gain margin or will permit you to use a less-effective preamp or, at the 10-foot (3-meter) end of the range, you can even get by with no preamp. Six- to 10-foot dishes are universally available from dealers in TVRO satellite equipment, and prices have become very competitive. Such antennas are available with either manual or remotely actuated polar mounts that permit you to swing across the entire geostationary arc—a real plus if you want easy access to all the GOES satellites. Such mounts are typically designed to mount on a pipe set into the ground in concrete but accessories for roof mounting are also available. If you buy a new antenna, get a quote on the system without the TVRO feed system or LNA. These items are of no use in our application and you should not be paying for them. Be aggressive in looking for the best price and you can save a great deal of money. Purchase of a used dish and mounting hardware from someone who has become disenchanted with satellite TV is still another option. Commercial surplus dishes are also available in some areas.

The primary disadvantage of large dishes is their narrow beamwidth. They must be aimed accurately—and if a satellite begins to drift near the end of its

Figure 2.7—Unlike TVRO satellite dishes, a WEFAX antenna need not seem particularly impressive. The difference in visual impact between this 4-foot WEFAX antenna and a typical 10- to 12-foot TVRO dish is considerable. The dish can be further blended into the background with the judicious use of paint if your living situation demands a low profile.

operational service, it is possible to get periods of the day during which the signal is degraded because the satellite drifts out of the antenna pattern.

Feed-Horn Design

The simplest approach to picking up the RF energy reflected from the dish is a feed horn, which is simply a metallic cylinder, closed at the far end. The cylinder acts as a waveguide at these microwave frequencies, and the RF energy inside the horn is picked up by a small probe coming in from the side (see Figure 2.8A). All of the dimensions of the horn, including its length (A), diameter (B), the distance between the probe and the closed end of the horn (C) interact to some extent and are dependent, along with the length of the probe, on the free-space wavelength at F, the frequency of interest (1691 MHz). All of the design equations are contained in Figure 2.8A, and I will not repeat them here. At 1691 MHz, the free-space wavelength is 17.74 cm, and that determines the minimum and maximum diameter of the horn (Bmin and Bmax) that will support propagation of the microwave energy within the horn. Our horn should have a diameter between 10.37 and 13.54 cm.

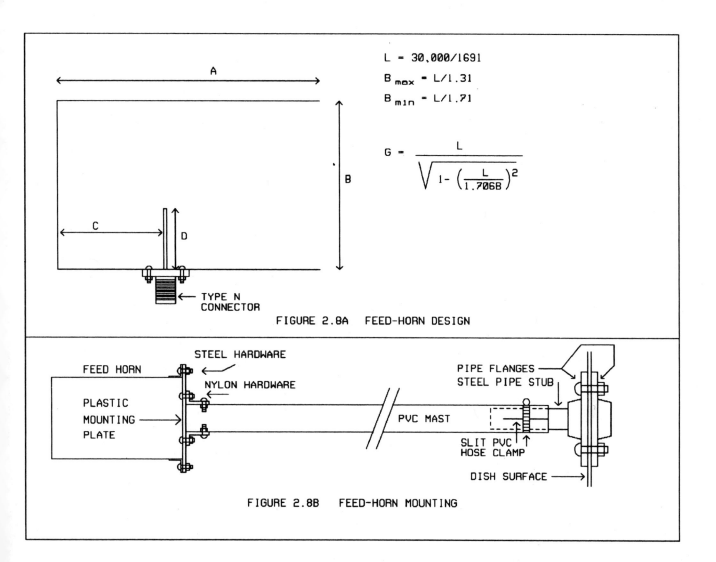

$$L = 30,000/1691$$

$$B_{max} = L/1.31$$

$$B_{min} = L/1.71$$

$$G = \frac{L}{\sqrt{1 - \left(\frac{L}{1.706B}\right)^2}}$$

FIGURE 2.8A   FEED-HORN DESIGN

FIGURE 2.8B   FEED-HORN MOUNTING

Once the microwave radiation is inside the horn, its wavelength changes to a so-called guide wavelength (G), which is dependent on both L and B. G is fairly complex to calculate. We need to know it to evaluate several other factors in the design of the horn, including the position of the probe relative to the closed end of the horn (C), the length of the pickup probe (D), and even the range of total length for the horn (A). Figure 2.9 is a nomogram that will let you quickly determine both G and C (the probe position) for any horn diameter in the permissible range.

You can custom fabricate your own horn from large-diameter tubing stock, but a perfectly suitable horn can be constructed from a 26-ounce (737-g) coffee can. These cans have a diameter of 12.7 cm which is just fine for dimension B, and results in a guide wavelength (G) of 31.5 cm (from Figure 2.9). The useful length of a horn of this type ranges from 0.5G to 1.5G, and the coffee can, with a length of 16.5 cm, exceeds the minimum length of 0.5G (15.25 cm) and is useful as-is.

Although the design formulas in Figure 2-8A allow you to determine if *any* particular can is usable and what its proper dimensions should be, some readers are reluctant to get involved with the math. For their benefit, I've written a simple BASIC program, HORN.BAS. You can download this program from the FILES area of the *Weather Satellite Handbook* BBS (see p 9-10). The program is written in generic BASIC and should run without a hitch on most systems. This program can be used to select a can size and determine the optimum dimensions for any frequency of interest to you.

HORN.BAS provides an interesting way to experiment with the variables in horn design. The program gives you the limits (in terms of horn diameter) for your chosen frequency, as well as an estimate of the range of horn lengths. The range is *only an estimate*, as it depends very much on the final diameter you use. In Table 2.1, you'll note that the horn length becomes unwieldy and impractical as you approach the lower limits of the horn diameter.

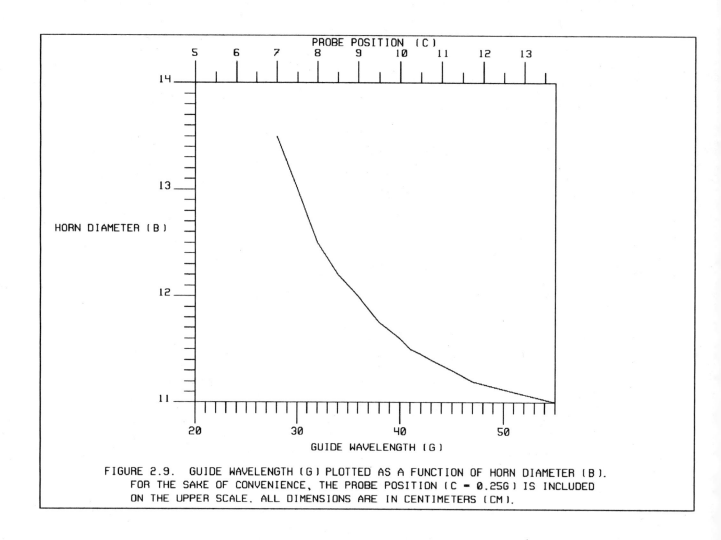

FIGURE 2.9. GUIDE WAVELENGTH (G) PLOTTED AS A FUNCTION OF HORN DIAMETER (B). FOR THE SAKE OF CONVENIENCE, THE PROBE POSITION (C = 0.25G) IS INCLUDED ON THE UPPER SCALE. ALL DIMENSIONS ARE IN CENTIMETERS (CM).

The coffee-can feed horn described shown in Figure 2.8A uses a type N connector. Although this is the best connector choice, some preamps use an SMA connector for the input. You can build the horn using other connector types. I strongly suggest, however, that you *never* use an SO-239 UHF connector. Stick to using a type N, SMA or BNC connector.

The Probe

In theory, the length of the pickup probe is dependent to some degree on its diameter. In practice, as long as the probe ranges from 1/8 inch (3 mm) to 3/16 inch (8 mm) in diameter, a length of 0.2L (3.6 cm) is about right. The probe should be positioned about 0.25G from the closed end of the horn; with our coffee can, this distance turns out to be about 7.7 cm.

Horn Construction

The description that follows will use dimensions based on the coffee-can horn. If your horn has different dimensions, use the design equations (Figure 2.8A) and the nomogram (Figure 2.9), or Table 2.1, to determine the critical dimensions, substituting them in the following description as needed.

Start construction by preparing the probe. The probe is constructed of a 3.6-cm (1.4-inch) piece of 5/32-inch (4-mm) brass tubing available at most hobby shops. Slip the brass tube over the center pin of a Type N female coax connector (UG-58A/U) and solder the tubing in place.

Measure 7.7 cm (3 inches) from the bottom (closed end) of the coffee can and use a 5/8-inch (1.6-cm) chassis punch to make a hole for the N connector/probe. Figure 2.8A shows the N connector mounted with screws, but that will invite problems with a poor ground to the body of the horn, particularly in the event of corrosion. It is best to place the probe through the hole in the wall of the horn and solder the connector into place. To do this, you should place a wire brush in an electric drill and clean the paint from the surface of the can around the hole, heat the bare metal with a propane torch, and tin the area with solder. If the body of the N connector is silver plated, the connector can now be soldered in place using the

Table 2.1

Horn Diam.	10.0	10.1	10.2	10.3	10.4	10.5	10.6	10.7	10.8	10.9
Probe Pos.	—	—	—	—	—	32.1	22.9	18.8	16.4	14.8
Min. Length	—	—	—	—	—	64.2	45.8	37.7	32.9	29.6
Max. Length	—	—	—	—	—	192.5	137.4	113.0	98.6	88.8

Horn Diam.	11.0	11.1	11.2	11.3	11.4	11.5	11.6	11.7	11.8	11.9
Probe Pos.	13.6	12.7	11.9	11.3	10.8	10.4	10.0	9.7	9.4	9.1
Min. Length	27.2	25.3	23.9	22.7	21.7	20.8	20.0	19.4	18.8	18.3
Max. Length	81.6	76.1	71.7	68.0	65.0	62.3	60.1	58.2	56.3	54.7

Horn Diam.	12.0	12.1	12.2	12.3	12.4	12.5	12.6	12.7	12.8	12.9
Probe Pos.	8.9	8.7	8.5	8.3	8.1	8.0	7.9	7.7	7.6	7.5
Min. Length	17.8	17.4	17.0	16.6	16.3	16.0	15.7	15.5	15.2	15.0
Max. Length	53.3	52.1	50.9	49.8	48.9	48.0	47.1	46.4	45.6	45.0

Horn Diam.	13.0	13.1	13.2	13.3	13.4	13.5	13.6	13.7	13.8	13.9
Probe Pos.	7.4	7.3	7.2	7.1	7.0	—	—	—	—	—
Min. Length	14.8	14.6	14.4	14.2	14.1	—	—	—	—	—
Max. Length	44.4	43.8	43.2	42.7	42.2	—	—	—	—	—

Refer to Figure 2.8—1691-MHz feed-horn dimensions as a function of horn diameter (B) over a range of 10.5 (minimum diameter) to 13.4 (maximum diameter) centimeters. Probe position (C), minimum horn length, and maximum horn length are shown. The design values are computed using the equations in Fig 2.8A, and all dimensions are in centimeters. Probe length (D) is 3.55 cm in all cases. This table permits selection of a suitable can or other metal cylinder if a 26-ounce coffee can is not available.

---

torch. Some of the new connector plating materials will not take solder. If this is the case with your connector, use the wire brush or a file to scrape off the plating on the mounting flange of the connector to expose the brass underneath. Tin the exposed brass and then solder the connector into place.

We are going to mount four small, brass, right-angle brackets equidistant around the outside of the open end of the horn. Such brackets are commonly available in hardware stores, or they can be fabricated from brass or steel stock. With the mounting areas marked, clean off the paint from each area with a wire brush and solder the brackets into place. Now go back over all the soldered areas with a wire brush to clean off any excess rosin, mask the N connector and probe with tape, and give the entire inside and outside of the can a coat of rustproof primer followed by several coats of a good exterior enamel to protect the can from corrosion.

So far, we have avoided having any unnecessary metal protruding into the horn; that is highly desirable to avoid distorting the delicate balance of RF fields on which the proper operation of the horn depends. We must now mount the horn without having any metal, including screws and other hardware, near the horn opening. All of the RF energy reflected by the dish must pass through the open end of the horn; even small amounts of metal can absorb a sig-

nificant portion of this energy and needlessly reduce gain. To mount the horn, we need a square of acrylic or Lucite plastic, or unclad G-10 fiberglass board material just slightly larger than the diameter of the horn. We will also need 3 or 4 nylon angle brackets and some no. 4 nylon hardware, all of which should be obtainable from your local hobby shop.

We will mount the plate to a piece of PVC plumbing pipe that will be used as a support to hold the horn at the proper distance from the center of the dish (Figure 2.8B). At your local hardware store or plumbing supply shop, purchase two metal pipe flanges and a short (1-foot) stub of galvanized pipe that will screw into one of the flanges. A 1-inch diameter pipe is about right, but you should choose a pipe diameter that is a slide fit into an available diameter of PVC tubing, so make your final choice of galvanized pipe and matching flanges while checking against the available PVC pipe. Unless you have a very large dish, approximately 8 feet of PVC pipe will be required. While you are at it, purchase a PVC coupler for your pipe size, and a bottle of PVC cement.

Cut off a 12-inch length of the PVC pipe. Take the short piece of PVC pipe and mount 3 or 4 nylon angle brackets flush with one end of the pipe using nylon hardware. Place the horn, face down, on the plastic plate and drill holes to match the mounting brackets around the mouth of the horn. Now, take the plate and, placing the PVC pipe with the nylon brackets at the center of the plastic plate, mark and drill holes to match the nylon brackets. Use nylon hardware to secure the plastic plate to the end of the 12-inch PVC pipe, then mount the horn to the plate using steel hardware. Figure 2.8B shows this overall arrangement. This figure shows the steel brackets bolted to the wall of the horn, but a better solution is to solder them in place. Now use the PVC cement to secure the coupling section at the free end of the pipe and set the assembly aside to dry.

To proceed with final mounting of the horn we need to know the focal length of the dish. Ideally, we want the horn to be positioned so the RF waves from the dish come to a focus just inside the open end of the horn. The focal length (A) can be calculated using the formula:

$$A = \frac{(0.5 \times D)^2}{4 \times Y}$$

where D is the dish diameter and Y is the depth of the dish at its center. To measure Y, lay a board or other straight edge across the face of the dish and measure from the center of the dish to the center of the board edge in contact with the rim of the dish. Either American or metric units can be used in the calculation, but don't mix American and metric units!

Temporarily insert the remaining PVC pipe into the pipe coupling attached to the horn assembly and measure along the PVC pipe to achieve a length equal to the focal length (A) minus 5 inches. If A = 24 inches, for example, you would measure a total length from the horn mounting plate of 19 inches. Remove the pipe from the coupling and cut it at the marked point with a hacksaw. Use the hacksaw to cut several 6-inch long slits in the far end of the PVC pipe. Now, use the PVC cement to insert the uncut end of the pipe into the coupling on the horn assembly and let the joint cure. Bolt the two pipe flanges, back to back, on both sides of the center of the dish and install the pipe stub (galvanized) so that it protrudes from the front center of the dish. Slide a stainless-steel hose clamp over the slit end of the PVC pipe, and slide the PVC over the galvanized pipe at the center of the dish. Twist the PVC horn mast so the probe is vertical, relative to the ground, and tighten the clamp so that the end of the PVC pipe mast is 3 inches from the dish surface.

If the PVC mast supporting the horn is not too long, guying of the horn may not be required. Simply connect a length of low-loss RG-8 coax (with a type N plug) to the connector on the horn, and run the line to the edge of the dish, or tape it along the length of the PVC mast. If the mast is quite long, or you'll be mounting a preamp right at the horn feed point, guy the horn using nylon fishing line from the corners of the horn mounting plate to the periphery of the dish.

Final Setup

Once your equipment is ready, use the information in Chapter 8 to point the antenna as accurately as possible in both azimuth and elevation. Listen for the satellite signal when the transmission schedule indicates the desired satellite to be active, and make minor adjustments in both azimuth and elevation to optimize the signal strength. Once you have achieved the best signal with antenna adjustments, loosen the clamp at the bottom of the PVC mast and slide the mast in and out on the pipe stub to further improve the signal. This latter step will optimize the position of the horn relative to the dish focal point. Prior to tightening the clamp, rotate the mast in either direction to optimize the polarization. When no further improvement is obtained, tighten the clamp and go inside to enjoy your new WEFAX installation!

COAXIAL TRANSMISSION LINES AND CONNECTORS

It is a fact of life that we must move RF signals around between antennas and various pieces of equipment; for that, we need coaxial cable transmission lines and the matching connectors. Losses are inevitable in RF cables and we need to minimize such losses whenever possible. Losses are a function of several factors including frequency, cable diameter, braid density, dielectric material (the insulating material—usually white—between the braid and the center conductor), and the diameter of the cable.

Operating Frequency

In a given type and grade of cable, losses will rise dramatically with operating frequency. This means that S-band losses will be considerably higher than comparable losses in the 137-MHz VHF range. This translates to the need for shorter runs of cable and the use of higher-quality cable for S-band.

Braid

The conductive braid beneath the outer plastic jacket of the cable is critical. Ideally this sheath should be solid metal and, indeed, the lowest-loss cables have such solid jackets, but the cables are expensive and inflexible. When the jacket is braided fine wire, as it is with most cable, braid density becomes critical. Bargain cables will have a low braid density: in some, you can even see the dielectric through the braid! A good cable will have very dense braid, and the fine copper wires that make it up will be plated to reduce corrosion. More-expensive cables will have a double thickness of braid, but we do not usually have to go to that extreme if we are careful in other aspects of cable selection.

Dielectric

The most commonly available cables have a dielectric of solid polyethylene. Still-better cables will have a foam polyethylene dielectric; these exhibit lower losses. Such cables are often designated as "foam," or carry an F in place of the normal R in the cable designation. Air can be used as a dielectric to achieve even lower losses. Teflon dielectric has a higher loss than polyethylene dielectric, but it also has a higher melting point. So, Teflon cables can carry more power than similar cables using a polyethylene dielectric. Air- and Teflon-dielectric cables are expensive.

Cable Diameter

All other factors being equal, losses are higher in small-diameter cables. The two generally available cables for 52 ohms impedance (the value we will use throughout most of the station) are type 8 (0.405 inch diameter) and type 58 (0.195 inch). The standard designation for the 8 cable is RG-8 with the foam types sometimes labeled F-8. The type 58 cables are represented by RG-58 with F-58 occasionally used for the foam version. Type 58 cable is suitable for short runs between equipment at VHF, but for any cable runs over 50 feet long you should use type 8. A good-quality type 8 foam will support runs of up to 10 feet at S-band, and you should not use type 58 at these frequencies for any runs over a few inches!

A great many manufacturers make coaxial cables, and each makes a number of grades. Although price is not an infallible guide, a higher price usually equates with better materials, quality control and lower losses. Avoid bargain CB-type cables as they are likely to exhibit high losses even at VHF and are totally unsuited for S-band use. In the balance between cost and quality, I recommend Belden 8214 for your type 8 cable needs, and Belden 8420 for type 58. In general, try to obtain cables of at least this quality to avoid unanticipated problems with your system.

Connectors

Like cables, connectors can generate both losses and instability. Constant-impedance connectors do not disrupt the impedance of a given line and will improve performance over connectors that are not matched, creating an impedance "bump" in the line with every connector set!

The most-readily available connectors are the so-called UHF series. The basic plug (male cable mounting connector) is the PL-259 designed for direct use with type 8 cable, and with type 58 and 59 (70-ohm) cable with the use of screw-in adapter sleeves. The matching chassis-mounting female connector is the SO-239. Threaded barrel connectors are available for connecting two PL-259 equipped cables. These connectors are not constant impedance types, and are marginally useful at VHF and completely unsuited for S-band.

A much better connector for VHF use, particularly with type 58 cables, is the BNC series. The military designation for the cable mounting plug is UG-88U. The matching flange mounted female connector, used for preamp and receiver inputs, is the UG-447/U. The UG-625/U is similar except that it uses thread mounting to the chassis wall. This is the connector of choice for your VHF connections using type 58 cables. These connectors can be used at S-band in a pinch, but regular use should be avoided.

The premier VHF connector and the most affordable option for S-band is the Type N series of connectors:

UG-21B/U	Type N plug for 8-series cable
UG-536B/U	Type N plug for 58-series cables
UG-58A/U	Type N flange mounting connector (receiver and preamp inputs)
UG-30/U	Type N double female connector (bulkhead mounting, weatherproof enclosures, etc).

No matter which connectors you employ, proper installation is a must for the best service and weatherproof integrity in the case of the N series cables. Manufacturers' data sheets cover proper installation as do publications such as *The ARRL Antenna Book* and *The ARRL Handbook*. All outdoor connections should be wrapped with plastic electrical tape, sealed with the putty-like sealer available at most outlets selling TV or other antenna components, and wrapped again with plastic electrical tape. Corrosion of connectors and moisture in transmission lines are the two major enemies of a trouble-free antenna system!

OPTIONS

If you have the misfortune of living in a situation that does not permit you to erect a permanent antenna system, you can still enjoy this hobby with the exercise of a little ingenuity. Let's assume for a moment that you live in a subdivision that does not permit antenna systems or you rent from a landlord that will not permit you to install an antenna. One solution would be to use the VHF Yagi, mounted on a short pole to make it easy to handle. With coax running back into the house, you can simply track the satellite manually from the backyard while a simple digital timer is used to activate the station recorder to tape the pass while you are working the antenna. With a digital watch and your tracking data on a 3 × 5 card, you can log the entire pass with no fuss, taking the antenna back inside when it is not in use. The neighbors may think you are a bit odd, but the technique works just as well as the most-elaborate antenna system. With the addition of a pair of headphones you can even track Meteor satellites with the simple linear Yagi, simply rotating the antenna to maximize the signal as you track the satellite!

Similar approaches can be used with a collapsible S-band dish, and it will even work when used from an apartment balcony for some satellites and/or passes.

If you live in an urban apartment with no balcony you still have options. The key here is to make your receiving system completely portable. A typical receiver will operate from a 12-V dc battery pack, and battery-operated stereo recorders are available. If you have access to the roof of your building, you can track and record a pass with no difficulty. Even if the roof is off limits, a neighborhood park or the parking lot of a shopping center can supply a site with enough open sky for effective tracking. With enough ingenuity, you can enjoy this hobby in almost any environment. It may not be as convenient as having everything permanently installed, but you will be able to enjoy the pictures that you do receive. If you have the interest, don't let a restrictive situation inhibit your creativity—it can be done!

Chapter 3

Weather-Satellite Receivers

INTRODUCTION TO VHF RECEIVERS

The most important link to that vehicle in space is your station receiver. You cannot compromise here. For, unless the receiver meets certain minimum requirements, you may be able to hear the satellite with a strong and consistent signal, but you'll *not* be able to obtain a satisfactory picture. To understand this seeming paradox, it's necessary to understand some aspects of the RF formats employed in weather-satellite transmissions of potential interest to amateurs.

All of the weather-satellite transmissions in which you are likely to be interested are transmitted as FM signals, and all polar-orbiter transmissions are made in the 137- to 138-MHz frequency range. Our basic FM receiver is made up of a series of circuit modules along the lines of the block diagram shown in Figure 3.1. In order to make a rational choice in terms of receiver selection, we need to know a bit about how each of these circuit elements functions. If you are interested in delving in depth into receiver design, chapters in a current edition of the *ARRL Handbook* provide comprehensive coverage. The following discussion, while brief, touches on some of the major factors that must be taken into consideration.

RF Preamplifier/Mixer

The VHF signals we receive from the weather satellites are quite weak. That shouldn't be surprising once you know that the satellite transmitter is rated at about 5 watts output (enough power to light a flashlight bulb), and the satellite's antenna has little gain. The job of the receiver RF amplifier and mixer stages is to provide some gain and selectivity before the signal is converted to a lower frequency for additional amplification and detection.

The gain and noise-figure characteristics of the receiver front end are two of the primary factors that determine how effective the receiver is. Because the satellite signal is quite weak, you might think that the gain of the RF amplifier stages (how much they amplify the incoming signal) is the primary concern, but this is not the case. Gain is secondary to front-end noise figure.

All electronic devices—including the first RF amplifier stage of our receiver—generate random noise. The noise generated by the first amplifier stage will be amplified (along with the desired signal) by all later stages in the receiver. If the first amplifier stage generates a lot of noise, we may never be able to hear the satellite signal, no matter how much additional amplification the receiver provides. The first amplifier stage needs to provide some amplification to overcome losses in the first mixer stage, but it must do so while adding the smallest amount of random noise to the signal—hence, our worship of low-noise front-end circuits.

In the late 60s and early 70s, typical tube-type amplifiers could provide front-end noise figures of 3 dB or higher. Very expensive tubes (do you remember the venerable 417A?) might push this figure down to around 2 dB, but circuit design and adjustment were tricky. Early bipolar transistor amplifiers could perform as well as some of the better tubes, but were easily overloaded by strong signals. The introduction of *f*ield *e*ffect *t*ransistors (FETs) was a major advance. *M*etal *o*xide FETs (MOSFETs) can provide a 2.5-dB noise figure with no fuss, and specially designed *j*unction FETs (JFETs) are capable of 2-dB noise figures with little potential for overloading. FET technology has advanced steadily and the latest designs, fabricated using *g*allium *ars*enide technology (GaAsFETs) easily achieve noise figures below 1 dB!

Although an overall reduction from 3 dB to 1 dB may seem small, remember that the decibel scale is logarithmic. An amplifier with a 3-dB noise figure limits the weak-signal performance of the receiver. The same amplifier, with a noise figure of less than 1 dB, is limited by external noise sources, either man-made or by the natural RF emission of the sun or other stars in our galaxy or beyond! Therefore, GaAsFETs are the devices of choice for the first RF amplifier stage. Succeeding amplifier stages (a receiver rarely has more than two stages of RF amplification) and the mixer also contribute noise, but the effect of this noise

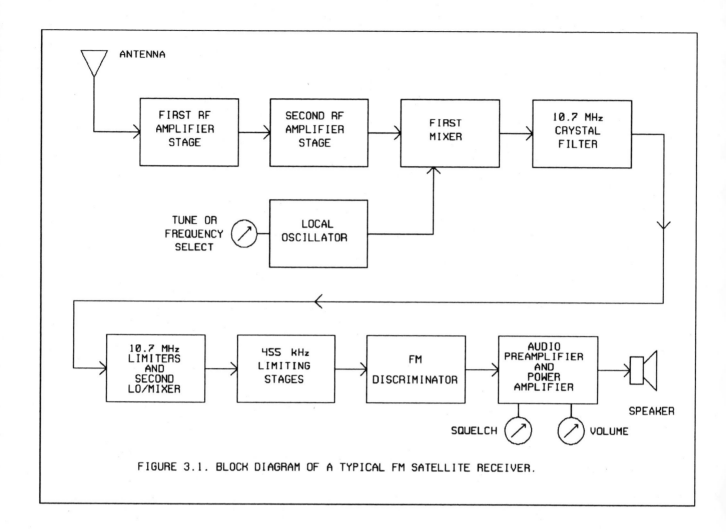

FIGURE 3.1. BLOCK DIAGRAM OF A TYPICAL FM SATELLITE RECEIVER.

is negligible provided the first stage has sufficient gain to—in effect—set the system noise figure.

Note, too, that signal losses in transmission lines prior to the first RF stage have the effect of *adding* to the effective noise figure of the front end. If we have a 1-dB noise-figure RF stage fed by a transmission line with 3 dB of signal loss, our effective front-end noise figure is 4 dB! Many of the newer receivers boast the use of a GaAsFET front end, but we cannot realize the full potential of the device if we precede it with excessive transmission-line losses. If we place a low-noise GaAsFET stage *at the antenna*, we can overcome the problem of transmission-line loss entirely.

The total gain of the RF front end should only be that required to effectively set the system noise figure and overcome any losses in the first mixer stage. Too much gain can overload the mixer, creating spurious products that degrade receiver performance. Excessive gain can also lead to the potential for overload and receiver desensing from unwanted signals outside of the desired frequency range. This is a major problem in most metropolitan areas due to the concentration of all sorts of RF-emitting devices. If you have a very

hot external preamplifier and relatively low transmission-line losses, you may actually have to reduce the gain of the receiver first-amplifier stage to avoid the overload problem. If gain is not easily altered, you may find it worthwhile to introduce some additional cable losses, or use an input attenuator to keep signal levels below the overload threshold.

Local Oscillator/Mixer

The mixer stage that follows the first RF amplifier(s) is actually a frequency converter. It mixes (or combines) the incoming signal with an RF signal from the local oscillator (LO) to produce outputs at several frequencies, including the LO frequency (L), the signal frequency (S), and two additional frequencies, S + L and S – L. In most basic receiver designs, we want to convert the signal to a lower frequency for effective amplification and filtering: an *intermediate frequency*, or IF. The S, L, and S + L frequency components are suppressed with tuned circuits that are adjusted to pass the S – L component with minimal losses. In simple receiver designs, the first IF is typically at 10.7 MHz.

If we want to convert a 137.5-MHz signal to 10.7 MHz, we can accomplish the job with either one of two LO frequencies: 126.8 MHz (137.5 – 126.8 = 10.7), or 148.2 MHz (148.2 – 137.5 = 10.7). An LO that is 10.7 MHz below the signal frequency (*low-side* LO injection) is usually chosen for circuit simplicity, but there is no reason why *high-side* injection (L = 10.7 MHz above S) cannot be used.

The LO signal thus determines what frequency is processed by the receiver. There are three basic ways to generate the LO signal:

- Use a tunable or variable-frequency oscillator (VFO)
- Use a crystal oscillator
- Generate the signal by digital synthesis

The VFO approach can provide continuous tuning over a specified frequency range, but at VHF, it is difficult to design a VFO that will be stable enough to provide reproducible calibration of the tuning range. A crystal oscillator can provide the needed stability, but requires the use of a different crystal for each frequency (or channel) we want to receive. The most advanced approach is to synthesize the LO signal digitally, providing the ability to tune a given frequency range in discrete steps, depending upon the design of the synthesizer. Most of the synthesizer circuits we'll encounter can provide tuning in 5-kHz steps.

IF Circuits

The receiver IF stages serve two roles. The first is to provide gain: the amplification of the signal prior to detection. In modern receivers, the IF stages consist of integrated circuits (ICs) that can provide very high gain (often greater than 100 dB) in the form of a simple chip. Most basic receivers actually have two blocks of IF circuits. The *first IF* is typically at 10.7 MHz. That is followed by a second mixer and crystal LO (often integrated into the 10.7-MHz IF-amplifier chip) that converts the signal to a 455-kHz *second IF* that provides still more signal gain. Very high gain in the IF stages is necessary because FM detection requires that we eliminate all of the amplitude variations in the incoming signal. The high-gain IF circuits accomplish this with *limiting* amplifier circuits, where all signals within a wide range will have a constant amplitude by the time the signals reach the FM detector circuits.

A second important role for the IF is to provide filtering of the signal. The IF stages can be looked upon as providing a frequency window, passing signals within a certain frequency range or *bandwidth* while providing significant rejection of signals outside of that range. The reasons for IF filtering are two-fold: to reject other unwanted signals that fall near the

10.7-MHz (or 455-kHz) IF, and to minimize the effects of noise.

The information in an FM signal is carried in the form of deviations from the basic signal frequency— hence the name *frequency modulation*. If the IF bandwidth is too narrow to accommodate the deviation (frequency swing) of the signal, some of the signal is clipped off by the IF filters and the signal will become distorted and generally unusable. On the other hand, if the IF is too broad, it will pass unwanted signals and noise that degrade the desired signal. Ideally, we want the IF to be wide enough to pass the signal with full deviation, with some additional allowance for errors in LO injection and changes in the signal frequency caused by Doppler shift.

An ideal satellite receiver has an IF bandwidth of about 40 kHz. The bandwidth is typically set by crystal or ceramic filters at the input of the 10.7-MHz IF stages, and by ceramic filters or simple tuned circuits at the 455-kHz stages.

Detector/Audio Stages

There are many ways to convert the varying signal frequency to a varying voltage that can be amplified to drive a speaker or other devices. Generally, the circuits that perform this job are known as *discriminators* or *ratio detectors*. In modern receivers, there is a trend toward the use of phase-locked loop (PLL) discriminators because of the excellent linearity typical of such circuits. The typical PLL discriminator is followed by a low-noise audio preamplifier and a power amplifier (usually a single IC) stage to drive a speaker.

Not shown in Figure 3.1, but present in most receiver designs, is provision for squelch control of the audio output. When no signal is present, the noise output of the receiver can be very distracting. The job of the squelch circuits is to shut off the audio output when no signal is present, and to turn on the audio stages when a signal appears. Squelch circuits can operate by detecting the presence of the signal either directly or indirectly by means of decreasing noise when a signal is present. The threshold point for squelch operation is usually adjustable by means of a squelch control.

Frequency Coverage

Our frequency range of interest is fairly narrow, with all polar-orbiter transmissions confined to the 137- to 138-MHz range. US TIROS/NOAA polar-orbiter transmissions are made on frequencies of 137.50 and 137.62 MHz. Each of the two operational satellites that are ideally in service at any one time use one of these frequencies. NOAA-10, for example, uses 137.50, while NOAA-11 uses 137.62. Each satellite has backup capability on the alternate frequency should a primary transmitter failure occur. Russian

Meteor/COSMOS weather-satellite transmissions occur on a wider variety of frequencies. In the past, 137.15 and 137.30 MHz were the two prime frequencies and there still appears to be an operational Meteor satellite, and often more than one, on 137.30 MHz most of the time. In recent years, 137.85 MHz has been used quite regularly, and 137.40 MHz has recently been placed into service as I prepare this edition. In any case, your receiver must cover 137.50 and 137.62 MHz; 137.30 and 137.85 MHz are desirable options.

Reception of geostationary WEFAX transmissions on 1691 and 1694.5 MHz (the latter used only by the European METEOSAT) is usually accomplished by using a converter ahead of the basic VHF FM satellite receiver. Such converters are designed so that the desired signal comes out at one of the "standard" VHF satellite frequencies (usually 137.50 MHz).

Receiver Bandwidth Considerations

Receiver bandwidth turns out to be one of the biggest hurdles to overcome since all of the various satellites use deviation values that are significantly higher than those employed for standard FM voice links. The biggest market for FM receivers (if we omit FM broadcast and TV sound) is for various kinds of scanners operating in the police and public-service bands. These transmissions typically deviate a maximum of ±7.5 kHz, and the receivers usually have a 15-kHz IF bandwidth. Unfortunately, if you tally up the values for signal deviation for a satellite such as the TIROS/NOAA series (±18 kHz) and Doppler shift (±3 kHz), you end up with a required bandwidth in excess of 40 kHz!

A signal from one of the polar orbiters, received on a typical scanner with 15-kHz bandwidth, will be severely distorted and won't produce a usable signal, though an unmodulated carrier sounds fine. Although deviation levels of the Meteor polar orbiters and the geostationary GOES and METEOSAT satellite are lower than those of the TIROS/NOAA satellites, their signals are still too wide for satisfactory reception with a stock 15-kHz-bandwidth receiver.

Receiver bandwidth is typically set by one or more filters in the IF chain. There is almost always a crystal or ceramic filter at the 10.7 MHz first IF, and there is often another filter (usually ceramic) at the 455-kHz second IF. If the receiver employs a crystal or ceramic filter at 10.7 MHz, the IF bandwidth can often be widened by simply replacing the original filter with an inexpensive 30-kHz crystal filter. A cheap 30-kHz filter has very wide skirts, and you'll end up with a receiver with sufficient bandwidth to do the job. In contrast, a good 30-kHz filter has steep skirts, and the receiver will still be too sharp for TIROS/NOAA service. If the

receiver also uses a 455-kHz ceramic filter, you may have a problem finding a replacement for it because wider-bandwidth ceramic filters at this IF are harder to come by. I have had good success in some instances by simply replacing the 455-kHz filter with a small-value (5- to 15-pF) coupling capacitor.

VHF RECEIVER OPTIONS

When selecting a receiver, new or used, its IF bandwidth should be your first consideration. In the sections that follow, I'll outline some general strategies for receiver selection and modification.

Commercial Satellite Receivers

Obviously, the simplest approach is to simply buy a receiver designed for satellite reception. Vanguard Labs has been one of the principal vendors for affordable satellite receivers over the years.

Figure 3.2—The Vanguard Labs Model FMR-250-11, an 11-channel, crystal-controlled VHF FM receiver. Many are still in service, but they are no longer available from Vanguard.

The FMR-250 (Figure 3.2) is an 11-channel, crystal-controlled receiver featuring dual conversion, a GaAs-FET front end, and a PLL FM detector. (Vanguard has replaced this receiver with its WEPIX 2000.) Volume, squelch, and frequency selector switches are on the front panel. The receiver has an internal speaker as well as a jack for hooking up an external speaker or audio network. The receiver comes supplied with a single crystal and operates from a 12- to 14-V dc supply. The latest receiver version comes complete with a wall-mount dc supply. (But you may want to use another supply; see **Receiver Modifications**.) A toggle switch mounted on the rear panel permits you to apply or bypass 12 V dc to the antenna transmission line to power remote preamps or converters. Although earlier models of this receiver had a 30-kHz-wide IF filter, the filter was too sharp for optimum polar-orbiter

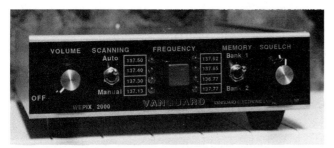

Figure 3.3—The Vanguard WEPIX 2000 synthesized receiver is housed in an enclosure the same size as that of the FMR-250 series, but the front panel includes a number of new displays and controls associated with the internal frequency synthesizer. Eight LED indicators mark the frequencies programmed by the manufacturer (137.13, 137.30, 137.40, 137.50, 137.62, 137.85, 136.77 and 137.77 MHz). A scanning mode toggle switch enables automatic sequential scanning of all channels, or can be set to provide manual stepping through the channels using the square push-button switch in the center of the panel. A memory bank toggle switch selects either Bank 1 (the frequencies noted above) or Bank 2 (eight channels at 5-kHz intervals, centered on 137.50 MHz). A 5-pin DIN socket on the rear apron carries the audio signal and TTL lines.

reception. The latest model features a slightly broader IF filter for optimum performance.

The WEPIX 2000 receiver (Figure 3.3) represents Vanguard's latest offering, and combines a number of features that bring this receiver very close to being optimum. A total of 16 frequencies, in two banks of 8, are programmed into the unit using an *erasable pro*grammable *read-only memory* (EPROM) chip. The first bank of 8 channels includes all active polar-orbiter frequencies. The second bank of 8 channels represent a series of channels, separated by 5 kHz, centered on 137.5 MHz. This bank is designed for use with a WEFAX downconverter, permitting fine-tuning of the IF to compensate for frequency errors due to thermal drift or other factors that would shift the output frequency of the downconverter. The active frequency is indicated by an associated LED. You can step through the channels manually, or use the receiver as an automatic scanner. Complete information on programming EPROMs for other frequencies is included. Up to 250 channels can be programmed into the EPROM; additional bit space is available for future control functions.

The receiver uses GaAsFETs for both the front end and mixer stages, assuring a low front-end noise figure. The receiver is a double-conversion design with a crystal IF filter and noise-operated squelch. A variable-gain output is available for driving your display system or recorder along with an independently adjustable speaker volume control. Two TTL outputs (one normally high and the other normally low) are provided for control of a tape deck or other external equipment.

Kits

Your most economical option is to build a receiver from a kit. The number of VHF FM receiver kits available is now quite limited, but fortunately Hamtronics still has several models, one of which is specifically broadband enough for satellite use (the RX75A). Such receivers are typically crystal controlled.

Converter/Communications Receiver

Another possibility, not often considered, is to use a converter ahead of a communications receiver equipped with an optional FM IF. If the FM option provides a sufficiently wide bandwidth, or can be modified to do so, this approach can be very effective. Older, HF-receiver VFOs didn't have a high degree of resetability, so it was difficult to return to and monitor a specific frequency. That problem doesn't exist with most modern HF receivers. You can modify a standard 2-meter converter to work into a 10-meter IF (add a new LO crystal and retune), or you can contact companies like Spectrum International and have them modify a converter for you. Hamtronics developed a converter that handles satellite-signal frequencies at the input and has an output in the 28- to 30-MHz range. Again, double-check the FM IF bandwidth specs: If the receiver IF bandwidth is too narrow, you'll get a strong but unusable signal!

Scanners

Now we come to the subject of scanner receivers! These radios come in three general configurations:

- First-generation crystal-controlled units.
- Programmable units covering the public-service bands.
- Wide-coverage programmable units.

Before discussing each category, a few general guidelines are in order. First, if you are going to buy a new unit, insist on seeing a schematic of the radio. In the case of the Radio Shack receivers, such a schematic is usually in the manual (although you may need a magnifying glass to read it). The purpose of this step is to see where the filters are located in the circuit. If you buy a used scanner, try to obtain a Sams Photofact sheet on it from your local electronics distributor. Try to stick with units that employ standard IFs (10.7 MHz and 455 kHz) as you'll have more luck in locating replacement filters to widen the IF bandwidth. (I personally avoid receivers that use IFs other than those.)

Figure 3.4—An example of a commercially modified scanning receiver offered by Software Systems Consulting. The receiver does an effective job and is economical, compared with other commercial options.

See if you can get the dealer to open the radio's case for you. (Most won't do so, but it's worth a try!) A peek inside the radio gives you a chance to assess the mechanical aspects of filter replacement.

Crystal-controlled scanners are by far the simplest to work with. Generally, the board layouts are more open, making it easier to get at the filters, and the VHF HI band (144-174 MHz) can usually be retuned to cover 137 MHz. Don't expect spectacular sensitivity, because the front ends are designed for broadband service. On the other hand, don't worry about it either: An external preamp will be needed anyway.

Programmable scanners designed strictly for public service use present two areas of concern. The first involves the tuning range. Most programmable scanners are strictly limited to 144- to 174-MHz coverage in the VHF HI range; entering a 137-MHz satellite frequency will only get you an error message. Two solutions to this problem exist: Set the receiver to a frequency that provides high-side LO injection for a satellite frequency, and retune the RF stages to peak at 137 MHz. Or, use a converter to move the 137- to 138-MHz satellite band coverage to a frequency range in the VHF LO band (usually 30 to 50 MHz). Hamtronics manufactures a converter (the CA144-6) to do the latter job.

The second problem concerns IF bandwidth. The PC boards of these receivers tend to have tight component layouts. Getting at and replacing filters can be a problem. The degree of miniaturization makes it imperative to locate equally small replacement filters; this can be difficult. Here is where examining the circuit and looking inside the radio can pay dividends prior to purchase!

The most impressive scanners of all are the wide-coverage units that represent the new generation of scanning receivers. At first, a receiver of this sort looks ideal because it covers the required frequency range, and the wideband FM mode (designed for TV sound and FM broadcast reception) is certainly wide enough for the satellite signal. Unfortunately, the wideband mode is *very* wide, resulting in a severe degradation of the signal-to-noise ratio.

A GaAsFET preamp at the VHF antenna will usually minimize noise and maximize signal to the point where a receiver is useful for polar-orbiter service, but its use for reception of geostationary WEFAX signals presents another problem. Even a GaAsFET preamp for 1691 MHz is significantly noisier than one used at 137 MHz, so the only solution for boosting the signal without introducing more receiver noise is to have higher antenna gain. You'll have relatively poor results in trying to use a small (4-ft [1.2-m]) dish with such a receiver, even though it delivers a satisfactory signal when used with a receiver having a 30- to 40-kHz IF bandwidth. The system will work if you use a GaAsFET preamp and a larger antenna (10 ft [3 m]) such as a TVRO dish, but all of this adds to the expense of your system, not to mention the required antenna space. All such problems would disappear if the narrow-band FM filters could be replaced with 30- to 40-kHz units.

Major receiver manufacturers such as ICOM, Kenwood, and Yaesu offer several synthesized receiver models that tune all of our VHF frequencies; some even tune the WEFAX S-band frequencies as well. Most of these receivers are in the $500 to $1000 price class. Although they're engineering marvels, they have the same IF-bandwidth limitations as the less-expensive programmable scanners. They are much larger and more impressive-looking than the simpler receivers noted earlier and may be cost-effective if you have other VHF and UHF monitoring interests beyond satellites. Dealers for these receiver lines can be found in the advertising pages of Amateur Radio magazines such as *QST*, *73 Magazine* and *CQ*, as well as more specialized publications such as *Popular Communications* or *Monitoring Times*. A good public or university library will have one or more of these magazines. Discount pricing can usually be obtained. If weather satellites are your primary interest, one of the commercial satellite receivers, matched for the proper IF bandwidth and optimized for our frequencies of interest, is a far better buy.

Surplus Receivers

In years past, using receivers retired from commercial FM service (police, fire, and other public-service uses) was a viable option. Older tube units featured 30-kHz IF bandwidths and are suitable for use with suitable frequency modification and retuning. Unfortunately, parts for such receivers are hard to obtain, and the use of an outboard preamplifier is a necessity to get a reasonable noise figure. Newer surplus units are solid state, but typically are designed for 15-kHz IF bandwidths, requiring additional modifications. Such receivers are of extremely high quality, however, and if you can obtain one inexpensively and feel confident in working with the conversion, the approach can be worthwhile.

Receiver Modifications

Regardless of which approach you take to acquiring the station receiver, a few additional modifications will be useful. The first concerns power supplies. A *very* well-regulated 12-V supply should be used. Ac-operated scanner supplies almost always have an unacceptable hum level. You may not notice the hum on voice reception, but it *will* show up in satellite pictures. Fortunately, most receivers have provisions for using external 12- to 14-V power supplies.

A second desirable modification is to provide a constant-amplitude audio tap from the receiver to your display system and tape recorder. This makes the video drive level independent of the receiver audio gain control and is quite handy. The easiest way to accomplish this is simply to tap off the top of the volume control (point furthest from ground) using a 0.1-µF disc or Mylar capacitor, and bring the line out to a phono jack at the rear of the receiver.

Preamplifiers

The use of an external preamplifier, preferably located at the antenna to reduce the effect of line losses, is highly desirable in any case, and a must in the case of the Zapper antenna described in Chapter 2. You'll also need a preamp if you plan to use the wideband FM option for the new generation of programmable scanners. Several vendors, including Spectrum International, Vanguard Labs, and Hamtronics (Figure 3.5) have suitable preamplifiers with some models equipped with weatherproof housings. Generally, these units are of two distinct types: JFET and GaAs-FET models. JFETs are very rugged and can provide noise figures of about 2 dB. The new generation of GaAsFETs can provide a noise floor below 1 dB. These are more delicate devices, however, that should not be exposed to intense RF fields.

Decisions! Decisions!

Given all the options for selecting a receiver, is there a best choice? Despite the diversity in the marketplace, if you are acquiring a receiver strictly for satellite use, the options are quite clear. One of the commercial satellite receivers is really your best buy. Any of these units has an almost unlimited service life. Repairs, if they are ever required, won't be costly. Considering

Figure 3.5—An external GaAsFET preamplifier provides the most effective single step in converting almost any receiver to state-of-the-art performance. This Hamtronics unit costs approximately $40. Locating the preamplifier as close to the antenna as possible helps overcome the problem of transmission-line losses.

the overall importance of the receiver, doing the job right should be your first objective. A crystal-controlled model is somewhat less expensive than a synthesized receiver, but only if you don't consider the final cost of all the crystals you might eventually add!

If finances are a problem, one of the Hamtronics kits will do a fine job, with you providing "sweat equity" (your labor for the construction of the unit). Remember that you'll have to add to the purchase price of the kit the cost of a cabinet, speaker, controls, and any crystals. Older model crystal-controlled scanners have very little resale value and it is often possible to find one for almost no cost. Such a unit will provide useful service with replacement of the IF filters and the addition of a modern GaAsFET preamplifier.

A wide-ranging scanner costs as much—or more—than a commercial receiver such as one of the Vanguard units, and won't perform as well. The more sophisticated receivers will easily double or triple your costs with no increase in performance. Unless you already have one, purchasing such a receiver is only justified if you have other VHF/UHF monitoring interests that require the added frequency coverage.

Trying It Out

Preliminary checks on the receiver can be made with a quarter-wave-vertical (20-in [51-cm]) antenna connected directly to the antenna jack. With such a setup, you should be able to clearly hear both TIROS/NOAA and Russian Meteor satellites during the best parts of a good pass, provided the receiver is outside (or in a wooden-frame building). There will probably be considerable fading and noise, but you *will* hear the satellite. Once the receiver is known to be working, it can be connected to the operational antenna/preamplifier system.

WEFAX DOWNCONVERTERS

WEFAX transmissions from the GOES, METEOSAT, and GMS satellite are made in the low-microwave region (S-band) at frequencies of 1691 and 1694.5 MHz. The most practical approach to receiving such signals is to use an S-band antenna in conjunction with a downconverter feeding the VHF FM station receiver. The job of the downconverter is to amplify the signal from the antenna, convert the 1690-MHz signals to the 136- to 138-MHz range, provide some VHF amplification to overcome mixer losses and have enough gain to eliminate the losses in the feed line between the converter and the VHF receiver.

When the Second Edition of the *Weather Satellite Handbook* went to press, amateurs were at the forefront of developing suitable converter options, despite the fact that early NASA publications on the new GOES system suggested that small ground stations were probably not practical. Now, we are in the position of having numerous off-the-shelf converters and receivers ready to do the job. Most amateurs today won't choose to embark on the relatively technical exercise of constructing their own converter. It's still important, however, to understand some of the major technical aspects of downconverter performance, if only to make an intelligent choice as to which system to purchase.

GOES WEFAX Link Analysis

When aerospace engineers plan a space communications link, they don't guess what transmitter power levels, antenna gain, and receiver sensitivity will be required to accomplish the desired communications. They engage in a very precise mathematical exercise known as a *link analysis* to predict the performance of the system. When we contemplate setting up a GOES receiving system, it's highly desirable to go through the same steps to determine whether we can make our system work with the equipment on hand, or within the limits of the funds that can be allocated to the project.

Aside from the outright purchase of a computer system for your station, WEFAX ground-station components will represent your single biggest investment. If you're going to plan and spend wisely, you need to take a very analytical approach to receiving system design. If you don't do so, you may find yourself with a relatively expensive collection of hardware that doesn't deliver the performance you require. If you prefer, you can skip the math and get to the bottom-line elements of system analysis. It's helpful, however, to know why you have to trade off on various elements of your system, so you might try to follow the basic elements of the math, even if it seems a little intimidating at first.

The basis of the link analysis is the comparison (under various conditions) of the power level of the satellite signal to the power level of the noise generated by your receiving system. The basic unit of power is the watt (W), but the power levels we'll be dealing with are so small, that it's more convenient to reference power levels to a much smaller unit—the milliwatt (mW), equal to 1/1000 of a watt. Our link analysis has three primary steps: the calculation of the power level of the satellite signal on the ground, the calculation of receiver noise power under various conditions, and, finally, how much antenna gain is required to deliver a useful signal with various receiver systems.

Satellite Ground-Signal Level

Path Loss

The satellite's signal power on the earth's surface is a function of two primary factors—the power of the signal radiated by the satellite and the *path loss* as the signal travels from the satellite to the ground station. The power radiated by the satellite—the *e*ffective *r*adiated *p*ower (ERP) is a function of many factors, including the output of the transmitter, coupling and transmission-line losses between the transmitter and the antenna, and the gain of the antenna. Fortunately, we can bypass all these complexities by simply looking at the manufacturing specifications that indicate an ERP of +54.4 dBm (+54.4 dB relative to 1 milliwatt), or approximately 260 W. The path loss depends on two factors: the operating frequency and the length of the path. Although the frequency is a constant (1691 MHz), the path length is variable, depending upon the look angle between the ground station and the satellite (see Chapter 8). A reasonable "average" value is 188 dB. Since this is a *loss*, we can compute the ground-signal level by subtracting the path loss from the ERP of the satellite:

+54.4 dBm (ERP) −188 dB (path loss) = −134.4 dBm

The power available from the satellite transmitter is obviously extremely minute, but to know what it means in practice depends on the relative noise power of our receiving system.

Receiver Noise Power

The noise power level at the input of our receiving system is critical. All electronic devices generate noise and, unless the level of power available from the signal exceeds the internal-noise power level by some reasonable margin, we won't receive a usable signal. The receiver noise power (TNL) can be calculated if we know two factors—the receiving system noise figure (NF_{dB}) and the system bandwidth (BW_{Hz}):

$$TNL_{dBm} = -174 + 10(\log(BW_{Hz})) + NF_{dB}$$

Because we want TNL to be as low as possible (so we can hear weak signals), we need a low noise figure, because noise figure *adds* to the noise power. Bandwidth is significant: The wider the bandwidth, the more noise comes through the system; hence, the noise power increases.

We could run through the receiver-noise-power equation with any combination of system noise figures and bandwidth values but for our purposes, a narrower range of values will highlight most system combinations. In terms of noise figure, we can examine two possible values: 3 dB and 1 dB. The 3-dB figure represents a realistic value for a suitably designed RF amplifier system using low-noise bipolar transistors, while 1 dB represents what can realistically be obtained using GaAsFET devices in the RF amplifier stage. For bandwidth, we'll look at two values: 30 kHz and 250 kHz. The 30-kHz figure represents a nominal IF bandwidth for a basic polar-orbiting-satellite receiver; 250 kHz is included because it represents a reasonable value for the wideband mode in the newer scanning receivers, and is also the minimum value required for reception of signals from the Japanese GMS satellite. If we use this range of values, we can calculate two values for each combination—the receiver noise power level (TNL) and the signal-to-noise ratio (S/N), obtained by comparing the TNL value with the available signal power level (– 134.4 dBm):

Noise Figure	System Bandwidth	TNL	S/N
1	30 kHz	–128.2 dBm	–6.2 dB
1	250 kHz	–119.0 dBm	–15.4 dB
3	30 kHz	–126.2 dBm	–8.2 dB
3	250 kHz	–117.0 dBm	–17.4 dB

As we might expect, a lower value for noise figure and/or bandwidth, does improve (lower) the receiver TNL value. The problem is, the signal is so weak, that even with a 1-dB noise figure and 30-kHz bandwidth, the signal ends up 6.2 dB *below* the noise-power threshold. With a 3-dB noise figure and a 250-kHz bandwidth, the signal is 17.4 dB below the noise threshold!

In order to display an image with minimal noise effects, we require a signal-to-noise ratio of at least +10 dB. Since our best-case scenario begins with a S/N of –6.2 dB, we obviously need at least a 16-dB boost in signal level because we've run out of options for reducing the receiver noise level. You might think that what we need is simple amplification, but addi-

tional amplifier stages amplify the signal power *and* the noise, and won't improve the relationship of the two. The needed boost in signal level—without "benefit" of additional noise—is available in the form of antenna gain!

Impact of Antenna Gain

Antenna options have been discussed in Chapter 2. Our calculation of ground-signal levels assumed no contribution from antenna gain. In practice, of course, you'll be using a gain antenna (and gain comes easily at 1691 MHz!), so the gain of your antenna can be *added* to the satellite-signal power level, thus improving the signal-to-noise ratio. Figure 3.7 plots the effect of antenna gain on system S/N for various combinations of receiver system noise figure and bandwidth. The nomogram covers antenna gain from 0 (with the same S/N values we calculated above) to 35 dBi—about equivalent to that of a 15-foot parabolic dish antenna.

A few levels on the S/N scale are worth noting. A level of 0 dB represents the point where the noise and desired signal have equal power levels. This point is somewhat academic, for although you could clearly hear the signal at 0 dB S/N, the signal would have far too much noise to be usable. The first really critical point on the S/N scale is the +10-dB signal level. This is the threshold level that would provide a noise-free display at the nominal maximum WEFAX power output (+54.4 dBm) from the transmitter. Under conditions of full-WEFAX power output, any combination of receiver noise figure, bandwidth, and antenna gain that is equal to, or above, this level will provide noise-free pictures. This signal level is an appropriate one for use with the GOES Central satellite (WEFAX only) under normal operating conditions. It is also a suitable level for use with the Japanese GMS satellite or the European METEOSAT satellite.

The major complication is the current operational practice with WEFAX from GOES. When the GOES program was first formulated, high-resolution transmission and WEFAX transmission were always scheduled in a non-overlapping manner, resulting in full-power output when the satellite was in the WEFAX mode. The satellites are now much busier, with additional systems, research operations, and time-intensive high-resolution scanning during severe-storm intervals. The result is that now, as a rule, the eastern and western satellites (engaged in active imaging) are now transmitting other data simultaneously with the WEFAX signal. The net result is that WEFAX signal levels are typically 6 dB *lower* than is the case with full-power WEFAX. Noise-free reception under such conditions thus requires a S/N of +16 dB, indicated by the WEFAX + VISSR threshold in Figure 3.7. If we want a system designed for worst-case GOES WEFAX reception, it would be best to use this threshold as your design target.

Putting It All Together

In addition to basic antenna gain values, Figure 3.7 also includes several specific antenna configurations, including parabolic dishes of 0.6 m (2 ft.), 1.2 m (4 ft), 1.8 m (6 ft), 2.4 m (8 ft), 3.0 m (10 ft), and 3.6 m (12 ft). Also included is a 45-element loop Yagi, an antenna that is popular because of its compact size and availability from several vendors.

Assuming a 30-kHz bandwidth for our basic satellite receiver, full-power WEFAX broadcasts could be received essentially noise-free with any of these antennas, although the 0.6-m dish and loop Yagi would yield very little *gain margin* (gain in addition to that required for noise-free display). Some gain margin is almost essential (see below), so of the two smaller antennas, the loop Yagi would be superior. A 1.2-m dish yields a solid gain margin (9.5 to 11.5 dB, depending upon noise figure) for full-power WEFAX reception and represents the practical minimum for reduced-power WEFAX operations. With a 3-dB noise figure, you'll have a slight bit of noise on the image, so the 1-dB system would be the conservative choice with this antenna. In effect, a conservative GOES or METEOSAT installation can be constructed around a 1.2- to 1.8-m dish (4 to 6 feet) with a 1-dB noise figure preferred for the smaller antenna.

Note that if we use a 250-kHz bandwidth, none of the antennas with less than 25 dB gain will yield noise-free display, even at full power WEFAX. To achieve an essentially noise-free display with such a receiver, a 3- to 3.6-m dish (10 to 12 ft) is required. This is the primary reason why I won't be promoting wideband receivers for GOES or METEOSAT reception.

If your station is located in the western Pacific, Asia, or the Australian, New Zealand area, your WEFAX objective will be the Japanese GMS satellite. This satellite transmits using a much wider FM deviation than either GOES or METEOSAT, and a receiver with 250-kHz bandwidth is absolutely required. The only saving grace in this situation is that WEFAX from GMS is almost always at full power, allowing you to design your station around the +10-dB signal threshold. Such an installation requires the use of at least a 1.8-m (6-ft) dish, and a system with a 1-dB noise-figure would be best at the low end of the antenna-size range.

Complications

All of this discussion to this point has assumed no other sources of signal loss or noise in the system. But there are transmission lines and connectors between the antenna feed point and the downconverter RF input. Losses in cables *add* to the system noise figure, so use of a minimum length of quality, low-loss cable is mandatory. Using improper connectors, or improperly installing good connectors, can create losses and further degrade system performance by creating input impedance mismatches that increase the noise figure of the first RF amplifier stage. Water or moisture in transmission lines or connectors radically increases losses, so the precautions noted in Chapter 2 are mandatory to avoid a gradual degradation of performance following initial installation. One of the major limitations of the loop-Yagi antenna designs is that the feed point on this antenna is very critical, yet it is exposed and susceptible to moisture pickup and corrosion—both of which quickly destroy the effectiveness of the antenna.

All of the previous calculations assume optimum matching of polarization at both ends of the circuit. In practice, this is most closely approximated with optimization of a linear feed horn as described in Chapter 2. If a circularly polarized feed is employed, system performance for any combination of factors will be degraded by 3 dB, but polarization won't have to be optimized. If polarization is 90 degrees out of phase (cross-polarized) with a linear feed, losses may exceed 20 dB. If this occurs, you would probably not even hear the satellite signal with any of the equipment combinations we have discussed.

Noticeable reductions in WEFAX signal levels are possible if the ground control station has not acquired the satellite properly with their own antennas, or the uplink power is reduced for any reason (WEFAX signal power is a linear function of the uplink power level with the GOES transponder). Small variations in signal level as a result of such factors are to be expected.

Also keep in mind that at low look angles (cases where the satellite is close to your horizon), result in longer path lengths (and hence greater pass loss) as well as greater potential for partial attenuation of the signal by foliage or other obstructions. Although normal weather variations have negligible effects, a very heavy rainfall can attenuate the signal to some degree.

The bottom line is that almost any additional factors can degrade the signal to some extent, while virtually nothing can happen to increase the signal level! Given this oppressive situation, it makes sense to design some gain margin into your system.

When evaluating converter performance, do not simply fix on the noise figure of the first RF stage in the system. I have constantly talked about *system* noise figure. The first RF stage can set the system noise figure only when one or more low-noise stages are used, providing sufficient gain to have the major influence in setting the system noise figure. The system noise figure will always be greater than the noise figure of the first RF stage. The difference can be minor in the case of a multistage, low-noise amplifier, but will

Figure 3.6—The Quorum Communications SDC1691 WEFAX downconverter. This is a state-of-the-art design with integral GaAsFET preamplifier and a thermal-stabilized LO chain that assures accurate frequency output over a wide range of ambient temperatures. An optional housing is available that permits the converter to be installed at the antenna for best performance.

increase if only a single low-noise, low-gain stage is employed. When in doubt, try to get the vendor to quote a system noise figure rather than the noise figure of the first RF stage.

Construction

Home construction of a downconverter is a project best suited to folks with a background in UHF and microwave construction. It is not that the projects are complex, for they can actually be deceptively simple. The real problem is that you typically have to have access to fairly sophisticated equipment in order to align and adjust your new creation. If you know what you are doing, extremely good results can be obtained at low cost, but you must be prepared to tinker.

Emiliani and Rhighini (*QST*, November 1980) describe a METEOSAT converter that has the virtue of using readily available, low-cost transistors, but the devices have relatively poor noise figures, and a number of stages of amplification are required because the gain of each stage is quite low. This particular converter was used with a 6-ft (2-m) dish to get the required gain margin. Those seriously interested in construction of a converter should follow technical articles in recent Amateur Radio publications such as *QST* and *The ARRL Handbook*, looking for descriptions of converters designed for use on the 1296- and 2304-MHz amateur bands because such designs can often be converted to 1691-MHz service. Those articles often delve into the design of the converters, something that is critically important if you are going to make modifications to published circuits.

Commercial Converters

Unlike the situation only a few years ago, a number of companies have affordable, off-the-shelf converters available for WEFAX service and new ones, including improved versions of existing models, will appear. Examples of commercial equipment with which I am familiar include the SDC-1691B from Quorum Communications (Figure 3.6), Spectrum International, Microwave Modules in the UK, and Wrasse Elektronik in Germany. These suppliers also have suitable preamplifiers available, including some using microwave GaAsFETs with noise figures below 1.5 dB! The SDC-1691B from Quorum Communications is an example of a fine state-of-the-art downconverter that combines a number of very desirable features at a reasonable price ($449). The unit features integral GaAsFET preamplification with a 1-dB noise figure. The LO chain is maintained within 1°C down to a temperature of –20°C, assuring that the output signal from the converter won't drift significantly over a wide ambient temperature range. This unit is also available with a weatherproof enclosure, thus simplifying installation.

Any downconverter is a significant investment, with prices typically in the $300 to $1000 range. Write suppliers for current model and pricing information as things change quickly in the dynamic world of precision electronics.

Weatherproofing

I noted earlier that the simplest approach was to mount the downconverter at the antenna. This re-

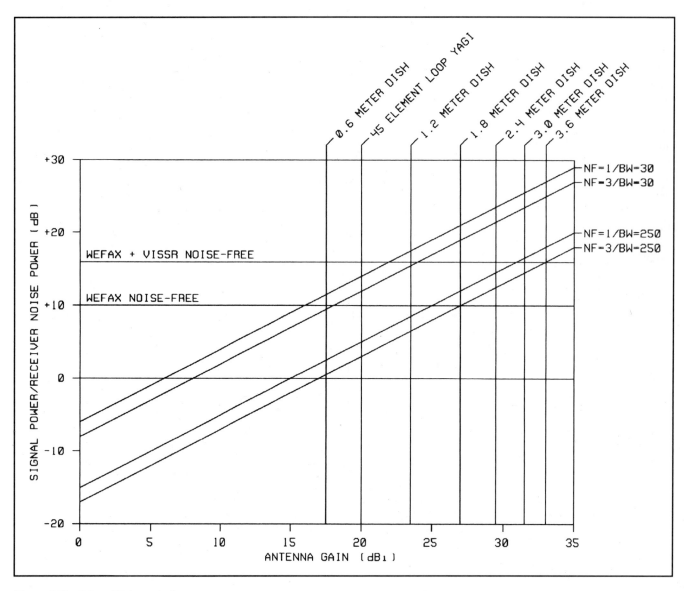

Figure 3.7—S-band link analysis nomogram.

quires some attention to weatherproofing, even with converters housed in supposedly weatherproof enclosures. If the converter is already weatherproof, it's still useful to provide a ventilated housing designed to keep the converter from direct exposure to the elements. Converters that are not protected by special housings require that you provide an external housing. Considering your investment in a downconverter, this job should be approached very carefully.

Industrial-grade housings, complete with weatherproof seals, are available and can be used in conjunction with weatherproof type N fittings to make a suitable enclosure. Other enclosures can be fabricated, but be extremely careful that your system is really watertight, for if water gets into a sealed enclosure, it can do more damage than if you used a ventilated housing designed to avoid direct water contact.

Cables should be routed with drip loops to avoid water running down cables and soaking connectors. The use of waterproof sealing compounds, designed for TVRO satellite installations may be helpful as well. Unless you can be absolutely certain of the integrity of your enclosure, periodic inspections are a must.

Frequency Drift

The output frequency of a converter is determined by its LO stages and the frequency output can be expected to change with temperature. This change will be small with a well-designed converter, but in the case of some converters, may exceed many tens of kilohertz with seasonal temperature changes at the antenna. The thermal stability of a converter is one of the things that you'll want to check when comparing equipment specifications from various vendors. A converter can be protected from high temperatures by shading the enclosure. Insulating the enclosure also limits thermal excursions.

If your converter drifts with temperature extremes, the receiver will require retuning unless the IF bandwidth is quite wide. If power can be routed to the antenna site, you may want to consider heating an insulated enclosure to some constant value above the highest normal ambient temperature. This will assure a constant output frequency, irrespective of the outside temperature. Such an installation may require the services of a qualified electrician to assure that you meet local code requirements, not to mention the common sense dictates of safety.

A subtle effect of temperature changes in some converters involves changes in the LO-injection level.

Some converters may begin to deliver a noisy signal at extremes of temperature. The LO chain in some converters may actually quit at temperature extremes. If your signal is consistently noisy at high or low temperatures, or disappears entirely, the need for thermal stabilization may be indicated.

Another alternative, if the antenna is not located too far from the building housing the receiver, is to use a low-noise preamp at the antenna and house the converter inside, away from the effects of weather and temperature extremes. Since most converters feature low-noise amplifiers at the 137.50-MHz IF, almost any length of transmission line can be used to connect the converter to the receiver elsewhere in the building.

Chapter 4

Video Formats and Display Systems

SPACECRAFT VIDEO FORMATS

Hands-on work with various satellite video signals is the fastest way to become acquainted with the details of the pictures transmitted by each type of spacecraft. The learning process will go a bit faster if you have some advance information about what you should be seeing. If your display system works well from the beginning, you're home free. If there are difficulties with your video circuits, knowing in advance what you should see will help you unravel your problems.

Subcarrier Modulation

As noted earlier, the signal you hear from your receiver is an audio tone with a frequency of 2400 Hz. While the tone will maintain the same frequency at all times, it will appear to be unsteady. This is because the amplitude or volume of the tone changes with different video values on the incoming signal. The tone has a definite maximum of peak value which represents white portions of the image. While the tone never quite declines to zero, it can drop to about 4% of the maximum amplitude—the level representing black areas in the image. Values between the black and white limits (4% and 100% amplitude respectively) represent various intermediate tonal values between black and white.

Displaying the image properly requires that we perform two tasks simultaneously. First, we must somehow convert the varying subcarrier (tone) amplitudes to their equivalent brightness values, a process known as *demodulation.* The goal is to change the display brightness in step with the amplitude changes in the subcarrier. In the case of a TV-like display, the brightness of the "spot" which is scanning the face of the tube must be altered. In a facsimile system, we must somehow vary the relative brightness of an image being printed on paper. All of these devices ultimately are controlled by a range of voltage that must somehow track the subcarrier amplitude changes.

The video circuits described in Chapter 5 provide one example of how this can be accomplished. A typical video circuit often involves some early *audio filter* stages, designed to pass the 2400 Hz subcarrier signal while rejecting noise and other anomalous signals at other frequencies. Such filtering can never be perfect, but it can noticeably improve the quality of the final image in less-than-perfect conditions.

Audio filtering stages are typically followed by one or more *detector* stages. More filtering follows, primarily to remove all traces of the original subcarrier signal. The final result is a dc voltage which varies in step with the changing subcarrier amplitude. Since any display circuit requires a certain range of voltage change to go from black to white, a video circuit also requires some gain control stages to assure that the output voltage has the right value to drive the final display circuits.

In the case of the computer interface described in the next chapter, we want a peak video voltage of +5 V for white, dropping to 0 V for black. The format diagrams in Figure 4.1 assume that we are looking at the output of a video circuit where the maximum output level represents white while the minimum level is black. The actual voltages vary with the type of display, but for our purposes we can assume a 0 to +5 V output level.

The spacecraft signal formats covered in this volume have a similar modulation scheme. Where they differ from one another is the timing involved in sending the video data. Our display system must precisely time the display process in step with the incoming image. If the timing is slightly off, the image may be tilted or skewed. If timing errors are greater, you will see no organized image but simply a hopeless melange of light and dark. In the sections which follow, we'll look at how the video information is organized according to each spacecraft type.

The diagrams in Figure 4.1 are analogous to what you might see if you looked at the output of a perfect satellite video circuit using a quality oscilloscope precisely in step with the incoming image data. To simplify

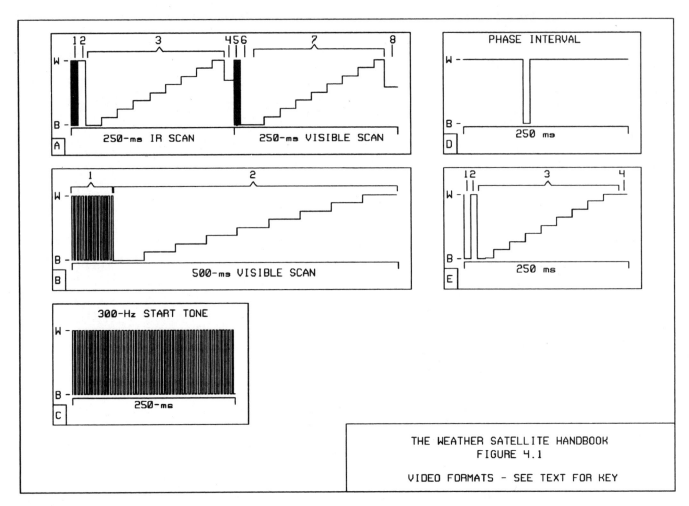

Figure 4.1—Video formats.

TIROS/NOAA APT FORMAT (FIGURE 4.1A)

the diagrams, the part of the signal that represents the image is assumed to be an 8-step grayscale, running from black at the left margin to white at the right.

TIROS/NOAA APT FORMAT (FIGURE 4.1A)

The multispectral radiometer of the operational NOAA spacecraft rotates at two revolutions per second, or 120 revolutions per minute. This rate of rotation is precisely controlled. One of the major tasks of any display system is to see that the timing on the display stays exactly in-step with the spinning radiometer on the distant spacecraft. One line of composite image data is transmitted with each revolution, so the basic line rate for these spacecraft is 2 lines/second, or 120 lines/minute (LPM). Since there are two lines each second and each second represents 1000 milliseconds (ms), the basic line of NOAA data is 500 ms long.

During the first half of each revolution, while the radiometer is actually scanning the Earth, the on-board computers insert the infra-red or IR data. During the second half of each revolution, when the scanning optics are pointed away from the Earth, the same computer inserts the visible light information into the video-data stream. Thus, each 500 ms line of image data is made up of two different types of data: IR information for the first 250 ms and visible light data for the second 250 ms. While the majority of each 250 ms interval is taken up with Earth scan data, there are a number of other distinctive elements that can help you know what kind of information you are seeing at any moment.

IR Sync Pulse: Each IR line begins with seven cycles where the subcarrier level swings from white to black and back to white. These seven transitions occur at a rate of 832 Hz. It is possible to design equipment that will detect this sequence of seven 832-Hz pulses, triggering the display of an IR line, but we do not use this pulse in any of our display systems. This train of pulses will appear as a fine series of vertical black and white bars if the sync pulse is visible on the display. It is the IR line sync pulse, and a similar visible light pulse that causes the very distinctive "tick-tock" sound on NOAA APT signals.

IR Pre-Earth Scan: Just before the sensor begins its scan of the Earth, there is a brief moment where it is viewing empty space. Since cold is represented by white in the IR data format, this view of space appears white and looks like a white stripe down the edge of an IR image. Once each minute, the spacecraft clock inserts minute markers into this pre-Earth scan zone of the image. These minute markers appear as thin, horizontal black markers going almost completely across the white, pre-Earth scan. These black markers provide a valuable time reference during the read-out of the image. The white pre-Earth scan itself is an infallible indicator that the video data to follow represents IR information.

IR Earth Scan: Most of the 250-ms IR line interval is taken up with the IR scan of the Earth's surface. Warmer objects appear darker than colder objects. If the display is set up for optimum rendition of visible light data, typical IR data is strongly biased toward the white end of the output voltage range, producing a low-contrast image of whites and light grays.

IR Telemetry: At the very end of the IR line is a zone devoted to telemetry information concerning various aspects of the satellite system. The information is coded as step-like changes in brightness, resulting in a strip down the right edge of the IR image made up of little grayscale step increments.

Visible Sync Pulse: The visible light segment immediately follows the IR line sequence. Like the IR line, it begins with a seven-pulse sync sequence. These pulses occur at a 1040 Hz rate so that equipment designed to respond to either IR or visible sync data. Since the visible light pulse rate is 1040 Hz, compared to 832 Hz for IR, the visible light pulses create a narrower sequence of vertical black and white stripes down the left edge of the image.

Visible Pre-Earth Scan: Since this pre-Earth scan represents visible light, the narrow view of space appears black . White minute markers are inserted into the visible light pre-Earth scan interval.

Visible Light Earth Scan: Most of the 250 ms visible-light interval is taken up by the Earth-scan data. This information is equivalent to what the eye would see and it is much easier to interpret than the corresponding IR data. Clouds appear in varying shades of white, water is almost black, and ground features are shown in intermediate gray shades. The precise range of grayscale values will depend on the intensity and angle of the sunlight on the Earth below the spacecraft.

Visible-Light Telemetry: The visible light line ends with a telemetry window similar, but not identical, to the corresponding one at the end of the IR line.

The previous description assumes that we are dealing with a daylight pass where both IR and visible light data are available. At night you will encounter one of two possible scenarios. The visible light line

interval may be completely dark, or the spacecraft ground control stations may direct the on-board computer to insert data from another IR channel (in which case both line segments will contain IR data). Since the IR sensors are sensitive to different regions of the IR spectrum, the two views are different in appearance.

If the display system operates at 120 LPM, the IR and visible light data are displayed side-by-side. This is rarely a good option since the great difference in IR and visible light contrast makes it difficult to get good renditions of both line segments without elaborate video processing. Unless your application demands a comparison between IR and visible data, it is usually best to run the display at 240 LPM. At this rate, either the visible light or IR data is displayed. The other data segment is "blanked." This allows the entire display image area to be devoted to either visible light or IR data, thus greatly improving the resolution of the displayed image.

There are a few subtleties about the APT format that might eventually occur to you. The first concerns image orientation. Any image displayed has the sync pulse and pre-Earth scan to the left of the image data. If you are sharp-eyed, you will note that in some of the images in this book the sync and pre-Earth scan appear on the right! These images represent *ascending,* or south to north passes (see Chapter 8). During these passes, the original image was displayed with south at the top (start) and north at the bottom (end). Since most of us prefer to see familiar geographic features with north at the top, many display systems have the facility to invert such images so the display "looks right." When the image is inverted, the sync and pre-Earth intervals will be on the right instead of the left!

The second subtlety involves *geometric distortion.* Imagine for a moment that you are seated in a glass airplane. You can look to the left at the horizon, rotate your head until you are looking straight down, and then continue the movement until you are looking out to the horizon on the right. When looking straight down, features are easily recognized. The view out either side, however, is distorted by the foreshortening of the horizon.

Since the scanning radiometer is viewing the Earth in a similar manner, the image *should* be distorted toward both horizons. Try as you will, you will not see such distortion in the images shown here. You would have seen it with the old ITOS/NOAA satellite series (NOAA 2 through NOAA 5), but is is absent from TIROS/NOAA images. This is caused by the activity of that busy little on-board computer. By skipping the actual horizon and varying the rate at which the remaining image data is sent to the APT transmitter, the effects of geometric distortion are largely removed, leaving an image that looks like a simple photograph.

120 LPM METEOR FORMAT (FIGURE 4.1B)

After the complexities of the NOAA image format, the Meteor signal seems simple by comparison. Standard Meteor images are transmitted at a rate of 2 lines/second (120 LPM) so that each line is 500 ms long. The start of the line is marked by 13 very prominent white to black transitions that make up the Meteor line sync pulse. On a properly displayed Meteor image, this pulse train will appear as a series of vertical black and white bars along the left edge of the picture. Although the bars usually appear black and white, occasionally it will appear as if you can see the underlying cloud cover through the bar pattern. In any case, the sync pulse train is very prominent and easy to recognize. The remaining line interval is taken up with the visible light image data.

Until recently, the typical Meteor spacecraft was strictly a visible-light system. Internal switching would turn the transmitter off when light levels dropped below a critical threshold. It was useless to listen at night for one of these spacecraft since the transmitters were always off! Descending passes on winter mornings are sometimes a bit surprising. The spacecraft can be above my horizon to the north, but the transmitter will still be off due to the low light levels from the polar landscape below. As the spacecraft moves south, light levels rise and at some point the transmitter will suddenly switch on. In such a case, the signal will seem to appear out of nowhere at full-quieting!

Now is is quite common to encounter Meteor spacecraft using the 120-LPM IR format. These images have the typical Meteor sync pulse, but look a little odd when displayed on your system. This is because the video values are inverted (clouds and cold objects are dark while warmer objects are white). To make the image conform to the more familiar IR format (cold = white and warm = black), the video must be inverted or complimented. In an analog system this requires an extra video-inversion stage (usually an op-amp). In digital systems the complimentation can be performed by mathematically manipulating the video data (see Chapter 10).

240 LPM COSMOS IMAGERY

As noted in Chapter 1, Russian COSMOS spacecraft are occasionally heard using a 240 LPM video format with standard subcarrier modulation. These spacecraft are in lower orbits than typical Meteor or TIROS/NOAA spacecraft, so the passes tend to be shorter and the area covered by an image is considerably less than you may be used to seeing. This smaller coverage area, together with the 240 LPM scanning rates, makes for spectacularly detailed pictures. Aside from the scanning rates, these transmissions can be identified by columns of numbers along one edge of the image.

Such spacecraft are only heard occasionally and they may tend to have short operational lifetimes. In any case, they are a real treat if you happen to catch one.

Occasionally you may hear 240 LPM transmissions that result in very unusual displays. These seem to be test vehicles for Russian radar satellites.

WEFAX SIGNAL FORMATS (FIGURE 4.1C-E)

TIROS/NOAA and Meteor transmissions are continuous, resulting in a strip of imagery built up throughout an entire pass. Any given picture from such a satellite simply represents a segment of the original transmission. In contrast, WEFAX transmissions are individual pictures or *frames* that typically require 200 seconds for display. The transmission format for a specific frame is rigidly organized in ways that make it easy to automate the display. The individual images are transmitted at a rate of 4 lines/second (240 LPM) using the standard subcarrier modulation format. Each WEFAX line is 250 ms in duration.

Start Tone (Figure 4.1C): A WEFAX image or frame begins with square-wave modulation of the subcarrier (between the black and white limits) at a frequency of 300 Hz. This produces a very distinctive tone from the receiver. If viewed on a display system it produces a total of 75 paired black and white vertical bars across the width of the display. This start tone interval is 5 seconds long in the case of the US GOES spacecraft, and 3 seconds in duration for the European METEOSAT and Japanese GMS spacecraft. The start tone can be used to automatically activate a display system at the start of a frame transmission.

Phasing Interval (Figure 4.1D): The start tone is followed by an interval where the subcarrier will be at the white level, interrupted by short (12 ms) pulses where the subcarrier drops to the black level. These black pulses occur 4 times each second and mark the point where the image lines begin in the upcoming picture. These pulses act as markers that allow the display system to get in step with the image—a process known as *phasing*. Some additional notes on phasing are included at the end of this chapter. The GOES phasing interval is 20 seconds long while the METEOSAT and GMS spacecraft have 5-second phasing intervals. Since the black phasing pulse is easily differentiated from the adjacent white-level data, it is relatively easy to implement an automatic phasing system.

Image Transmission (Figure 4.1E): Once the phasing interval is finished, the actual transmission of image data begins. There will be a narrow vertical black bar at the extreme left edge of the line (1), a narrow white-level framing interval immediately to the right (2), the image video data (3), and a narrow white border (4) to the extreme right end of the line. The nominal WEFAX frame consists of 800 lines. Since the lines are transmitted at a rate of 4 lines/second (240 LPM), the required transmission time is 200 seconds. Tropical East and West quadrants will be slightly shorter. Some weather charts will be a bit longer.

Stop Tone: When the frame is complete, the subcarrier is modulated between the black and white limits at a frequency of 450 Hz. Like the start tone, the stop tone is easily recognized. It's used, in automatic systems, to disable the display system at the end of the picture. The duration of the stop tone is about 5 seconds.

High-Frequency WEFAX Format

The characteristics of satellite video formats are useful in understanding facsimile transmissions on shortwave (high-frequency or HF) bands. However, there are some critical differences that you need to understand if you are going to deal with these images.

Subcarrier Modulation: AM subcarrier modulation can be used for satellite images because the spacecraft use FM modulation. The limiting action of the FM receiver means that variations in signal strength from the spacecraft will not result in variations in the peak level of the subcarrier. The HF or short-wave bands are very crowded. FM modulation of the signal would result in excessive bandwidth. As a result, HF radio transmissions are typically AM, using single-sideband (SSB).

SSB receivers do not feature limiting, so fading and other variations in the signal path can cause an AM subcarrier to vary in amplitude. To get around this problem, we can use FM modulation of the subcarrier. Instead of the constant frequency of 2400 Hz, the subcarrier frequency is varied to convey video information, while its amplitude is held constant. The subcarrier amplitude will be variable by the time it arrives at your receiver, but these variations have no meaning in terms of the image. It is changes in the *frequency* of the subcarrier that are significant. Virtually all HF fax services use a subcarrier frequency of 1500 Hz to represent black while 2300 Hz is used for white. Intermediate grayscale values are represented by frequencies between the 1500 and 2300 Hz limits.

Signal Format: The standard format for the transmission of weather charts and images on shortwave is similar to the satellite WEFAX format, with a few critical differences. Instead of 240 LPM, the HF line rate is 120 LPM. A 300-Hz start tone is used (5 seconds), consisting of modulation between the black (1500 Hz) and white (2300 Hz) limits. The start tone is followed by a 5-second phasing interval. In the case of HF WEFAX, the phase pulse is white (2300 Hz), rather than black (1500 Hz), against a black (1500 Hz) background. The length of the frame which follows is variable, depending upon the specific image being transmitted. It usu-ally varies from about 10 to 16 minutes. A 450-Hz stop tone (3 seconds) signals the end of the transmission.

Although the subcarrier modulation is different on HF, procedures for processing the data in software can be quite similar to those used for satellite WEFAX and 120-LPM METEOR transmissions.

DISPLAY SYSTEMS

There are three basic approaches to the display of weather satellite pictures: slow-scan CRT monitors, facsimile recorders, and various forms of scan conversion. Let's review each of these briefly.

CRT Display

Direct, cathode ray tube (CRT) display of the weather satellite formats represents one of the most easily visualized and easy-to-implement forms of image display. The operation of a CRT system has much in common with a conventional TV display. In a conventional TV set, an electron beam scans across the face of the tube at the proper horizontal rate. This creates *scanning line.* As the beam moves across the screen, the intensity of the beam varies. As a result, the intensity of the glow varies, too.

At the same time the beam is scanning horizontally, it is also moving vertically. Each line "written" on the face of the tube falls just below the previous line. The combination of vertical and horizontal scanning, in combination with the varying brightness of the trace, "paints" a complete picture. The type P4 phosphor on the face of a standard B&W TV tube has a very short "persistence," which means that it does not glow very long after the electron beam passes over a given point on the screen. The factor that gives the picture the illusion of permanence on the screen is the very fast rate at which the picture is painted. The face of the tube is scanned horizontally over 15000 times each second. A complete image forms every 1/60th of a second. The eye simply cannot perceive such rapid changes. The brain integrates all the information and interprets the face of the tube as containing a complete picture.

In principle, a CRT monitor works the same way. In the case of a WEFAX picture, the lines arrive at the rate of 4 each second over the 200 seconds required to create an 800-line picture. If we scan the tube horizontally at the 4-Hz rate, scan it vertically from top to bottom in 200 seconds, and cause the trace to lighten and darken in response to the variations in subcarrier amplitude, we can paint the WEFAX picture on the face of the tube.

Unfortunately, the picture information is painted on the tube slowly. As a result, we cannot see the entire image at one time. If we use a P4 tube, all we can see is the current line being scanned. If we were to use a

Figure 4.2—A home-brew CRT display system.

Figure 4.3—Display of a NOAA-7 visible-light image (August 17, 1983) showing hurricane Alicia making landfall on the US Gulf Coast.

tube with a long-persistence phosphor such as those used for radar displays (phosphor types such as the P7), we might be able to retain up to a few dozen or more lines on the screen. We can, however, take a time-exposure photograph of the screen over the 200-second frame period. The film, when processed, would contain the complete picture.

Monitors that work directly at satellite-scanning rates must involve photography in order to see the picture in its entirety. This is a major limitation of the CRT approach.

Even so, there are some very positive points in favor of the CRT approach. Satellites differ primarily in matters such as scanning rates. Since CRT scanning rates can be altered with simple circuit modifications, a monitor is easily constructed or altered to handle virtually any image format. It is this simplicity and flexibility that have made CRT displays quite popular among weather satellite enthusiasts *despite* the fact that you must photograph the screen. Your major chore, should you choose to take this approach, is acquiring specialized parts such as the CRT tube, high voltage power supply components and deflection yokes. Considerable latitude is possible and creative salvaging of parts from small-screen black and white TV sets will often suffice. An excellent example of a CRT display system can be found in Figure 4.2.

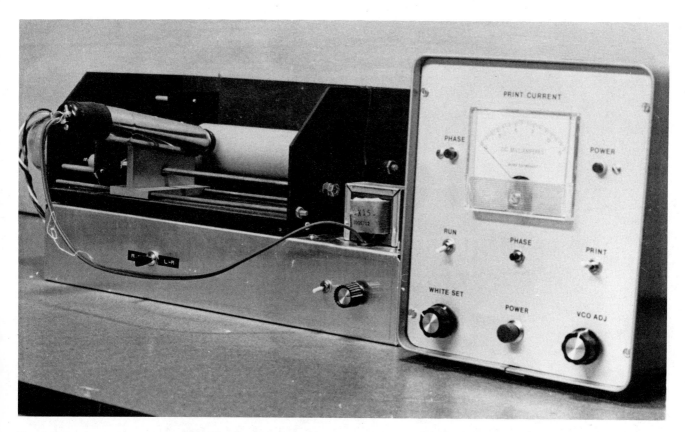

Figure 4.4—A home-built drum facsimile recorder constructed by the author. The control and drive electronics are housed in a metal cabinet and the various signals are routed to and from the drum and motor assembly with a multiconductor cable. The mechanical assembly includes precision synchronous motors for drum drive (240 r/min) and for driving a threaded rod that moves the printer carriage (40 r/min). This particular system is set up for direct printing on sheets of photographic enlarging paper so printing is via a light gun made up of an R1168 glow modulator tube and a lens assembly from the eyepiece of a microscope. The lens is required to ensure that the modulated light source is focused as a fine spot of light on the paper wrapped around the drum. The electronics assembly contains video circuits similar to those in Chapter 5, a high-voltage driver circuit (300 V maximum) for the glow modulator tube, a crystal-controlled 60-Hz frequency reference for drum speed control, a 60-Hz power amplifier and step-up transformer to provide the 110 V ac required by the drum motor, and a variety of circuits for control and phasing.

Facsimile (fax) Recorders

A fax recorder is a combination of electronic circuits and mechanical features that creates a direct print of the satellite image. The print can be created using light on sensitive photographic paper, producing a permanent image when processed. Other methods use various kinds of electrolytic or electrostatic recording papers. The major advantage of fax is the fact that you have a direct and immediate (or almost immediate) printout of the picture. The prints are typically large and extremely detailed. Unlike a photographic print, the fax recorder is limited to a single-size format due to the mechanical nature of the printing system. This mechanical factor also affects flexibility. The scanning rates in a fax recorder are determined by motor speeds, gears and mechanical linkages. Changing scanning rates is fairly complicated compared to the modest changes needed to achieve the same change with CRT systems.

While a CRT system is almost all electronic, a fax recorder involves a considerable amount of mechanical machinery. A home-built fax recorder is a tinkerer's delight. With everything working just right, a fax recorder can produce the best possible images with the least bother. However, a well-built CRT system can come very close with considerably less effort. The drum of the fax recorder must operate at a very precise speed. This requires the use of a synchronous motor driven by an audio power amplifier operating from a crystal reference source. Video circuits are similar to those in Chapter 5, but the output circuits will vary depending upon the type of recording paper used. Photographic paper or film will require the use of a

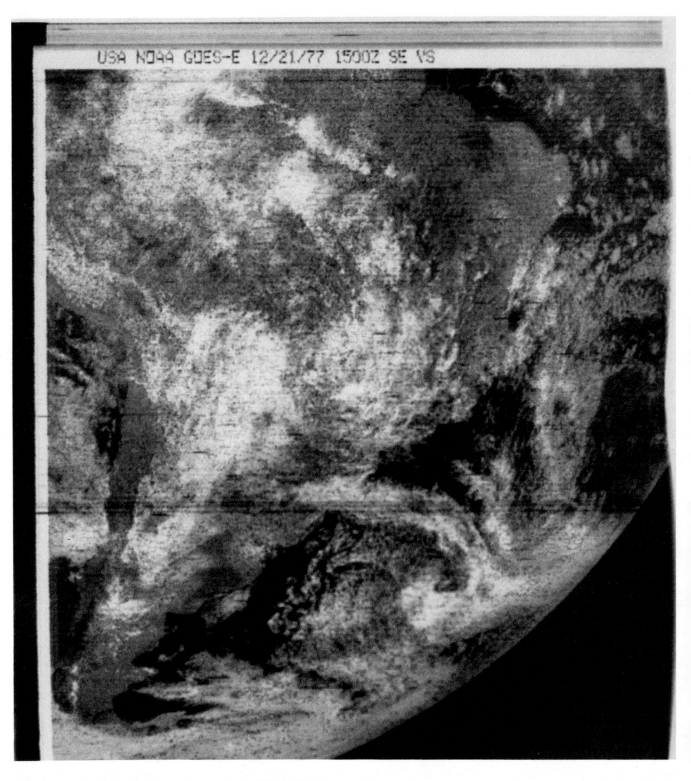

USA NOAA GOES-E 12/21/77 1500Z SE VS

Figure 4.5—A photographic fax printout of a GOES WEFAX frame. This image was printed using the fax recorder illustrated in Figure 4.4. The light source was a crater lamp with a lens system constructed from microscope optics. The system was set up to print directly onto photographic enlarging paper, producing a 7 × 7-inch image.

modulated light source in conjunction with a lens system. High-intensity light-emitting diodes (LEDs) might be suitable if sensitive photographic film is used as the recording medium. In such a case, the video circuits must provide maximum brightness on white and minimum brightness on black.

Photographic papers eliminate the need to process the film and then expose and process the final print. However, a brighter light source must be used. In either case, white subcarrier must generate minimum light output while black results in maximum lamp output. Because a photographic recorder uses light-sensitive paper or film, the recorder must operate in the dark until processing is complete.

Fax machines using electrostatic papers typically apply a modulated voltage (45 V for white up to about 300 V for black) to a wire stylus. Electrolytic papers are commonly used with blade and helix continuous-read-out recorders. In such designs, printing voltages in the 75 to 150 V range are employed.

The major problem in constructing and operating a fax recorder is the steadily rising cost of new synchronous motors. Obtaining a supply of the specialized papers can also be a headache. Figure 4.4 illustrates one of my drum-type fax recorders with a sample of its photographic output in Figure 4.5.

The Computer Connection

While all of the previous approaches to satellite image display will work, you won't find such projects in this edition. The days of CRT screen photographs and mechanical fax recorders are long gone—thanks to personal computers. To be more specific, it is not just computers, but the ability of computers to generate complex graphic images that have made all the difference.

The first generation of personal computers treated the monitor screen as simply a place to print text, with some primitive graphics thrown in as an extra. That was to change, however. Falling prices for computer memory chips and increasing computer speed and power combined to make it easier for computers to handle images of increasing complexity. The computer you buy at the local mall or discount store often has the capability to create images of photographic quality.

If you can find a way to transform weather-satellite data into something the computer can understand, it can display that image, analyze it, and process it in ways impossible with other systems. The computer can even store it for immediate recall at a later time. It can do all of this without using a single piece of fax or photographic paper—and the results will typically surpass anything you could accomplish with a CRT monitor or fax machine.

That same computer can also keep track of satellite orbits, tend your station when you aren't home, and

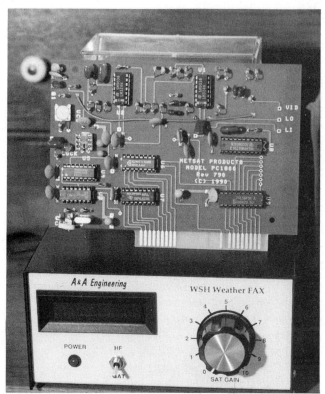

Figure 4.6—The key to using your computer to display satellite images is interface circuits that process the satellite video and provide timing signals to the computer. The two most common approaches to implementing such an interface are shown here. The upper unit is a plug-in interface card designed to use the accessory card slots available on most IBM-compatible computers. Such circuits can be quite simple, because the computer houses the card and provides the dc voltages required for operation. A slightly more complex, but much more flexible interface is shown below. This is the kit-built version of the stand-alone interface described in Chapter 5 with the optional LED level indicator. This unit will process both satellite and HF fax signals. It connects to your computer via the printer port.

receive and store pictures in your absence. What's more, it still runs all those educational programs for the kids and balances the checkbook! Given all of this, it is little wonder that the computer has replaced all other approaches to satellite image display.

When the first all-digital edition of the *Weather Satellite Handbook* (the 4th) went to press, a computer system capable of processing satellite imagery cost at least $3000. Just this morning, I strolled down the main street of my college town (East Lansing, Michigan) and saw a computer on sale at the local Radio Shack for $650. This computer had all the power, speed, and image display capability needed for satellite use—at about $\frac{1}{5}$ of the going price just a few years ago. If cost were the most critical item on my agenda, I could pick up a used computer and be in business for much less than the costs of building a CRT monitor or fax machine. If my budget was really tight, I could record satellite data on audio tape and use a computer at school or work to display the image!

In order for a computer to process and display the incoming video data from the spacecraft, it requires *interface circuits*. The interface processes the analog subcarrier signal and converts it to digital or numeric values. In addition to the interface, we also need to provide some precision timing information so the computer can properly assemble the incoming image. There are two basic approaches to implementing this interface, both of which are illustrated in Figure 4.6. The most popular approach is to put all the required circuits on a card designed to plug into an available card slot in the computer. Alternatively, the circuits can be incorporated into an external unit connected to the computer with a multiconductor cable. Both approaches have their merits and the question is addressed in the context of the WSH Satellite Interface design in Chapter 5.

If we are going to process and display images on a computer, it's useful to understand a few basic issues with respect to digital image display and storage. These are discussed in Chapter 6. While under-

Figure 4.7—Computer display of satellite images provides the convenience of fax, with immediate display of the picture. It requires no specialized supplies such as photographic papers or chemicals. In addition, the images can be processed to optimize the display and the pictures can be stored on disk for later viewing.

standing digital image basics is not a requirement for using the available software, knowing about the subject can help you select the best interface hardware and processing software.

Chapter 5

The WSH Satellite Interface

THE PROBLEM

Even with the decision made to focus on VGA-equipped IBM-compatible computers, there is still the question of how to implement a satellite interface unit. The most obvious approach, used by most of the vendors advertising in this edition, is to put all the necessary circuits on a card designed for installation inside the computer. This avoids the need for a cabinet and the computer system provides all the operating voltages. It is simple, convenient and popular. The available cards all work well and your decision, if you choose to buy one, will be based primarily on the price and the features that are supported in software.

I have nothing against such imaging cards. I even designed one shortly after the fourth edition of this book went to press. It was simpler than most, worked like a charm, and even won me some laptop computers in the Zenith Data Systems *Masters of Innovation IV* competition. I am pleased with the design, but you won't see it in this edition. The reason is simple. No imaging card can meet the criteria required for use in this *Handbook*. I want my designs to be simple and reproducible, modest in cost yet with state-of-the art performance, and as universal in application as possible. With a computer-based display system, this includes the ability to use it with the widest possible range of computers and the flexibility to serve as a basis for further experiments.

While my imaging card makes the grade with simplicity, cost, and performance, it fails the universal applications standard. There are a number of reasons and they apply to all the imaging card variations you will find:

The no-slot problem. To use an imaging card, you must have available slots in your computer. There are many excellent VGA-equipped compatibles that don't have slots. These include all the laptops and notebooks, the base models of the IBM PS/1, and several personal computers from other major manufacturers. They have the speed, memory, and display capabilities, but they don't have slots. No slot, no card!

The wrong slot problem. Most IBM PS/2 models employ the Microchannel bus. They take cards, but not cards designed for the standard PC/AT bus. Microchannel cards represent a more specialized market, they tend to cost more than PC/AT cards for a specific application, and there are few, if any, weather satellite interface cards available.

The empty slot problem. Even if your computer has them, many of today's systems come with three available slots and it doesn't take long for them to fill up with cards for other applications.

The too many computers problem. You may have more than one computer you would like to use to display pictures, or you may be in a situation where you don't have your own system and have to depend on a computer at work or school. Swapping an imaging card from system to system is a pain. It's also something I'd rather not do if the computer doesn't belong to me. I face this situation all the time when giving demonstrations and a number of folks in the educational field have the same problem.

The wrong computer problem. Although there are many practical reasons for settling on the IBM-compatible computer family for our display system, there are other computer families, including the Apple Macintosh and the Commodore Amiga. All of these computers feature image display capability. Unfortunately, there is little interface software or hardware for these systems, despite their many merits. Ideally, our IBM-compatible interface could be used with such systems, without the expensive "bridge" software and hardware normally required to run MS-DOS applications.

The anxiety problem. You will notice that there are essentially no kit options out there for imaging cards. The people who make the cards are concerned about liability in a society where very few people take responsibility for their own mistakes. If you build a kit but do something wrong that causes expensive damage to your computer, the kit supplier worries that your lawyer will fault their design, the components supplied, or the instructions—anything to cover the damages, not to mention the cost of the lawyer! It is safer to supply only wired and tested cards. In a similar vein, most people worry about building something that plugs into their computer. I probably have as much experience designing and building exotic hardware as anyone, and even I feel a twinge of anxiety when I throw the power switch on a new project!

THE SOLUTION

The answer is simple. We can have the same functionality as an imaging card, and avoid all the problems I've mentioned, *by moving the interface circuits outside the computer*! With a little bit of cleverness on our part, our external interface can "talk" to the computer through its parallel printer port—a gateway between the computer and the outside world that is present on virtually all computers now in use. This permits the interface to be used on systems without proper card slots, allows it to be quickly shifted between computers, and has the potential for interfacing to non-IBM-compatible systems. There is virtually no chance that you can harm your computer if you do something wrong, and the approach makes it easy to experiment with new ideas if you choose to do so.

The only disadvantage to this approach is the need for a housing for the unit, a power supply, and a cable to interconnect the interface with the computer. This is balanced out by a slightly simpler circuit for the external interface, since we can delete three chips normally required to interface with the computer's address/data bus. Even with a cabinet and power supply, the external interface is far more economical than any imaging card, yet it provides equal or better performance.

In this chapter I discuss the design and construction of the weather satellite interface. It ended up being one of the simplest and most economical designs I have ever presented, without any compromise in performance. In addition to displaying all the WEFAX and APT formats transmitted by the various operational satellites, it will also handle the signal format used for HF WEFAX transmissions, allowing you to view satellite images (and weather maps) with nothing more elaborate than a short-wave receiver and a simple wire antenna!

While it is perfectly feasible to build the interface from scratch, using perf-board and ordering up all the parts, most people appreciate more support in a construction project. A double-sided pc board with plated-through holes and solder masking is available, complete with a silk-screen overlay showing the placement of all parts. For those who don't want the hassle of finding the parts, the circuit board is available as part of a comprehensive parts kit, complete with a drilled, punched, and silk-screened cabinet and all necessary hardware. For those who want to minimize the effort required to get up and running, the interface is also available as a wired and tested unit. All of these options are available through A&A Engineering in California. A complete listing of the board and kit options is included at the end of this chapter.

A complete PC-compatible program (WSHFAX) is available for the interface. It is comprehensive enough to meet the needs of 90% or more of those interested in setting up a ground station. This software package is included in the kit and wired-and-tested interface packages at no extra cost. For those building from scratch or using just the printed circuit card, the software can be purchased separately for $25. The interface is essentially open-ended in terms of its performance capabilities and I encourage those who would like to develop alternative software packages. Chapter 6 is devoted primarily to working with interface software development, so there are no mysteries or propriety aspects to the design. You can develop your own software and I encourage developers to port their own software to the interface. You may choose to upgrade software in the future if you make significant additions to your computer system. Even so, the interface hardware will still continue to meet your needs, no matter how elaborate your computer system may become.

Finally, the design of the interface is such that it is an easy matter to interface it to computers outside of the PC-compatible family. There is almost no weather satellite hardware for these alternative computers and the WSH Interface can solve that problem. Although software development is different for these other computer families, more than enough information is included in Chapter 6 to encourage developers to write the needed code.

The remainder of this chapter will be devoted to the design, construction, and testing of the interface unit. Chapter 9 will contain a complete set of instructions for operation of the software package, as well as providing an example of software implementation that may be useful in guiding your own software development efforts.

CIRCUIT DESIGN

There are four basic circuit modules for an interface of this type: the AM (satellite) and FM (short-wave) video demodulators, a precision clock circuit, and the interface circuits required to get the clock and video data into the computer through the parallel printer port. This functional breakdown is ideal as the organizational structure for discussing the various circuit elements.

AM (Satellite) Demodulator (Figure 5.1)

The image signal from any of the weather satellites consists of an amplitude-modulated (AM) subcarrier (2400 Hz) that will vary in amplitude from maximum (100%) peak amplitude for white, down to 4% of peak

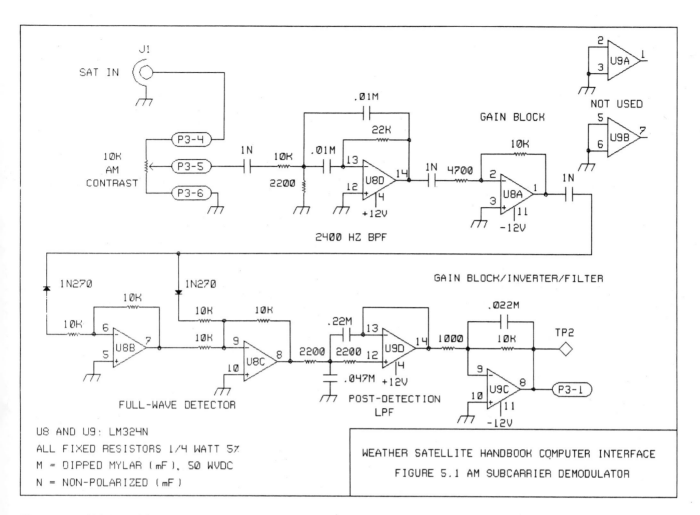

Figure 5.1—AM demodulator

amplitude for black. Intermediate grayscale levels are represented by peak amplitudes between the 4% and 100% limits. The AM video demodulator circuits are designed to convert this signal to a dc voltage that shifts from 0 V for black up to +5 V for white.

Satellite video signals, typically derived from the station receiver, are routed into the interface at the SAT jack (J1) on the rear apron of the interface enclosure. From that point the signal is routed off the board by way of connector P3 (pin 4) to the front-panel SAT CONTRAST control that is used to set the peak (white) level of the AM demodulator. The variable signal at the center-arm of this control is routed back to the circuit board through P3-5, where it is applied to the input of an active band-pass filter made up of U8D. This filter is designed to minimize the effect of extraneous noise on the input signal and has a center-frequency of 2400 Hz, a bandwidth of 1600 Hz, and unity voltage gain. The input filter is followed by a simple op-amp ac amplifier stage (U8A) that provides a voltage gain of 2 with the components shown.

The two remaining stages of U8 are used to implement a full-wave audio detector. U8B passes the negative-going components of the subcarrier signal and inverts these to a positive-going signal with no change in amplitude. The output of U8B is summed with the positive-going subcarrier signal component at the input of U8C. The output of U8C is a negative-going signal consisting of 4800-Hz signal peaks whose amplitude is proportional to the instantaneous peak-to-peak amplitude of the original 2400-Hz subcarrier signal.

All that must be done at this point is to filter the output of U8C to remove the 4800-Hz component remaining from full-wave detection. U9D is wired as a low-pass filter to accomplish the filtering function. U9C is an output buffer that accomplishes three functions: a voltage gain of 10 to overcome filter losses, inversion of the negative-going signal to a positive-going waveform, and some additional post-detection filtering. Assuming the front-panel AM CONTRAST control has been properly adjusted, the output of U9C, as measured at TP2, will vary from 0 V with black input,

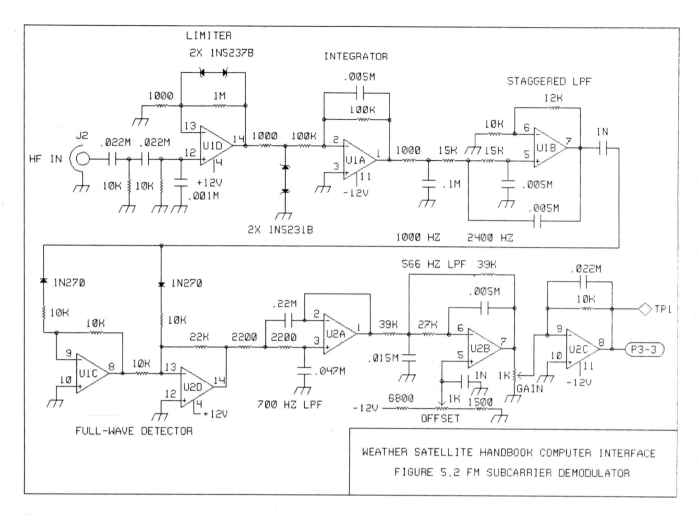

Figure 5.2—FM demodulator

FM HF WEFAX Demodulator (Figure 5.2)

In contrast to the AM modulation of the satellite signal subcarrier, HF or short-wave WEFAX transmissions employ FM modulation of the audio subcarrier. This minimizes the effects of signal level variations that are inevitable when receiving short-wave transmissions. At the transmitting station, the audio tone is varied from 1500 Hz for black areas of the image, to 2300 Hz for white image components. The FM demodulator must convert these subcarrier frequency variations to the same 0 V (black) to +5 V (white) range realized in the output of the AM demodulator.

Although we are dealing with audio frequencies here, as opposed to RF, the functional components of the FM demodulator are quite similar to those employed in FM receivers (see Chapter 3). We need limiting stages to remove any unintended AM modulation, a discriminator to convert frequency variations to amplitude changes, and a detector. We also need post-detection filters to remove the subcarrier products remaining after detection.

HF WEFAX signals are applied at the HF jack (J2) on the rear apron of the interface. Signals from the short-wave receiver consist of the frequency-modulated audio subcarrier (1500 to 2300 Hz) which also is varying in amplitude as a result of variable fading of the received signal. U1D is a two-stage audio limiter that clamps the FM signal to ±8 V using D1 and D2, then to ±5 V using D3 and D4. This provides good limiting over a wide range of input levels. The output of U1D is integrated in U1A and applied to a staggered low-pass filter (U1B) which functions as a slope detection FM discriminator.

U1B is designed to have a linear response between 1000 and 2400 Hz. The 2300 Hz white subcarrier is attenuated comparatively little in passing through the filter while the 1500 Hz (black) subcarrier is attenuated significantly. The output of U1B is thus effectively

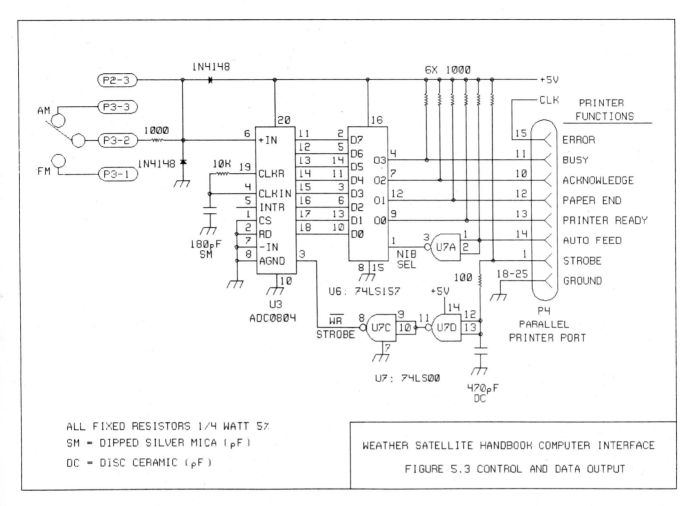

Figure 5.3—Control and data output

AM modulated at this point with black at a low level and white at a much higher amplitude. From this point on, the FM detector circuits have a great deal in common with the AM circuits previously discussed.

The slope-discriminator is followed by a full-wave detector (U1C and U2D) essentially identical to that employed in the AM circuits, except that the summing stage (U2D) has a gain of 2 in this circuit. Since the frequency products resulting from full- wave detection range from 3000 to 4600 Hz, in contrast to the 4800 Hz in the AM circuits, a more sophisticated post-detection filtering system is required. In this case, two low-pass stages are used. U2A comprises the first stage. It is essentially identical to the low-pass filter in the AM circuits. This is followed by a second low-pass stage made up of U2B. This stage also provides an adjustable dc offset by way of the OFFSET potentiometer to permit adjustment of the black level output. The gain and offset levels were specifically chosen to provide a white output should the HF signal fade.

The output buffer (U2C) provides signal inversion, adjustable gain through the GAIN potentiometer (Rg), and additional post-detection filtering. The output waveform is monitored at TP1. With 1500 Hz input, OFFSET is adjusted for 0 V output. With 2300 Hz input, GAIN is adjusted for +5 V. The output of the FM demodulator is routed to the FM side of the front-panel AM/FM selector switch (SW1) via P3-3.

Clock Circuits (Figure 5.4)

All program functions involving the input and display of satellite image data must be performed with extreme accuracy to get a high-quality picture. This is achieved by a precision clock reference (CLK) tied to a master crystal oscillator.

U4C and U4D comprise a TTL oscillator whose frequency is controlled by a 4.194304-MHz crystal. A series trimmer capacitor (C25) allows the oscillator to be trimmed for precisely this frequency, compensating for manufacturing tolerances in the crystal. U4A and

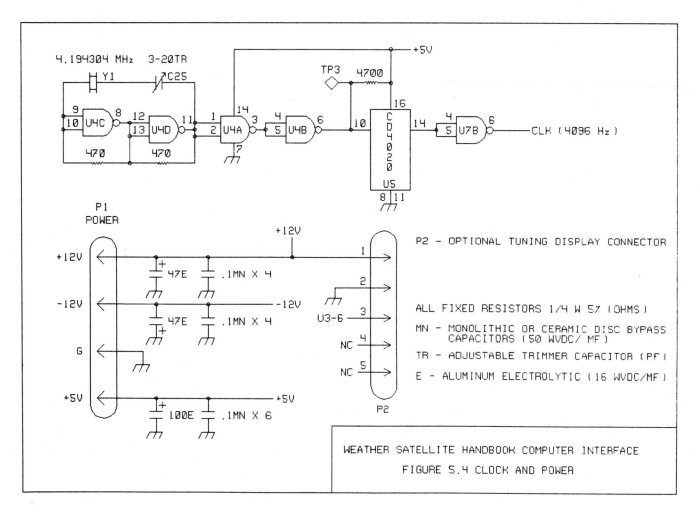

Figure 5.4—Clock and power supply connections.

U4B function as buffers, isolating the oscillator from the counter chip that follows. The oscillator frequency can be measured at TP3 to facilitate precision adjustment using a frequency counter. In practice, the oscillator is trimmed as closely as possible to 4.194304 MHz, since even an error of 100 Hz can result in a significant "slant" in the displayed images.

U5 is a multi-stage binary counter, which, wired as shown, provides a divide-by-1024 function. With an input frequency of 4.194304 MHz, the output at pin 14 will be 4096 Hz. This permits precision loading of up to 1024 pixels in a 250-ms WEFAX or APT line, or 2048 pixels for a 120-LPM format. These values are essentially the same as the theoretical resolution limit for the two formats (1200 and 2400 pixels/line, respectively). The 4096 Hz CLK signal is buffered by U7B to produce the master CLK reference signal.

Parallel Port Interfacing (Figure 5.3)

The CLK timing reference signal is routed to pin 15 of the DB-25 parallel port connector (P4) on the rear

apron of the interface. This timing signal can be read by the computer since this line is normally used to carry the ERROR status signal when a printer is connected to the port.

The analog 0 V to +5 V signals from the AM and FM demodulators must be converted to digital values in order for the computer to use the information for image display. The common line from the front-panel **AM/FM** switch (SW1) at P3-2 is routed to the input of U3, an 8-bit analog to digital (A/D) converter chip.

An A/D conversion is initiated by pulling pin 3 (WR) LOW and then returning it to a HIGH state. Depending upon the individual device, valid 8-bit output data will then be available in 40 to 70 ms. The computer will wait until CLK (ERROR) goes HIGH. At that point, the STROBE output at P4-1 will be pulled LOW. This line is buffered by two gates (U7C and U7D) so, when STROBE goes LOW, so does pin 3 of U3. The computer will allow a short interval for settling of the logic level at pin 3 (typically much less than 1 ms). At this point STROBE will be returned to a

HIGH that will also appear at pin 3 of U3. This starts the conversion process. The computer will then wait for the CLK (ERROR) line to go LOW—a period of 122 ms with a CLK frequency of 4096 Hz. This allows ample time for U3 to complete the conversion and assures that the output data is valid.

U3's output consists of eight data lines, allowing coding for one of 256 possible digital values over the input voltage range of 0 to +5 V. On some versions of the PC/AT-compatible printer port, it would be an easy matter to find a total of 8 input lines for the video data. Many of these options will not work on all versions of the parallel port, however. The most conservative approach is to use the five STATUS input lines that are implemented on all versions of the parallel port. We have already allocated one of these lines (ERROR) to the CLK line, leaving four available input lines.

U6 provides the means to let just four lines do the work of eight! U6 is a multiplexer that functions as a four-pole digital switch. When pin 1 of U6 is LOW, the four high-order data lines from U3 (D4-D7)—the *high nibble* of the video byte—are routed to the four output lines. If pin 1 is set HIGH, the four low-order data bits (D0-D3)—also known as the *low nibble*—are sent to the four output lines. The status of pin 1 of U6 is controlled by the AUTO FEED control output from the computer parallel port. AUTO FEED (P4-14) is buffered by one NAND gate (U7A), so a HIGH applied to AUTO FEED results in a LOW to pin 1, making the high nibble available to be read. AUTO FEED can then be set LOW, setting pin 1 HIGH and allowing the low nibble to be read. In effect, the computer reads the high nibble, sets AUTO FEED, reads the low nibble, and then resets AUTO FEED. With additional formatting of the two pieces of 4-bit data, the entire 8-bit video value is available for processing.

There are a few tedious complications in using the parallel port to read the clock and video values, but the computer can get the job done in short order. Chapter 6 contains sample code and a more elaborate discussion of all the required routines.

Power Supply Connections (Figure 5.4)

Three voltages, all relatively low current, are required for the interface. Plus and minus 12 V dc is required by the op-amps in the video circuits. Plus 5 V is needed by the TTL chips and A/D converter. There are many small ac-operated power supply modules that will provide the required voltages. One such unit is supplied with the kit and wired-and-tested interface options from A&A Engineering.

CONSTRUCTION

Although the interface can be constructed using perf-board and point-to-point soldered or wire-wrapped connections, the use of the commercial printed-circuit board (Figure 5.5) from A&A Engineering (see listing at the end of this chapter) makes the job very easy. A&A can also supply complete parts kits, as well as wired and tested units. The circuit board measures 4×6 inches (10.2×15.2 cm) and is double-sided with plated-through holes. The board is solder-masked on both sides and silk-screened with outlines and identifiers for all parts.

Although most readers will probably obtain the complete parts kit, I have included a complete parts list at the end of this chapter. If you are *not* going to use the circuit board, I suggest that you prepare several photocopies of the circuit diagrams and mark the connections, as you make them, with colored pencils. Once complete, I suggest that you wait a few days and then recheck your work, just to be sure that you did it right. If you are using the circuit board, all these worries are eliminated. Proper interconnections are assured if you mount the parts as indicated. From this point on, I will assume the use of the A&A board, although many of the comments that follow are applicable to perf-board construction as well.

The tools required to construct the project are minimal. This is especially true with the kit, since all the drilling and other cabinet work is already done. You will need small needlenosed pliers, diagonal wire cutters, and a soldering iron (20-30 W) with a small pointed or chisel tip. You will also need some small-diameter rosin-core solder. A wire-stripping tool will come in handy as well. A copy of the *ARRL Handbook* is an invaluable guide to proper construction, component labeling, and soldering techniques.

Clean solder connections are an essential element in a successful project. If you are uncertain about your abilities, contact a local radio amateur, electronics experimenter, or computer "hardware hacker" for assistance. Most will find the project irresistible and they can help you in many ways.

In one sense, construction of the main circuit board is quite simple—you stuff the board with parts, solder them in place, and proceed with final assembly. Since the main circuit board contains 95% of the actual interface circuitry, quality construction is critical. It can make the difference between a short period simply optimizing a few adjustments or an extended period of trouble-shooting. There are three simple strategies to assure a successful outcome: identification, orientation, and soldering!

Identification: Put the right part in the correct holes! Double and triple check to make sure that you have selected the proper component and the proper holes for mounting it.

Orientation: Orient the parts properly! Integrated circuits, diodes, and polarized capacitors have one proper orientation. Make sure it's right before solder-

Figure 5.5—Interior view of the interface unit. The front panel is to the left with the rear apron on the right. The small ac power supply module is located toward the top of the photograph with the ac power cable, switch, and fuse on the rear apron. The DB-25F parallel port connector and the satellite and HF input jacks mount to the PC board and protrude through cut-outs on the rear aprom. The **AM/FM** selector switch, **SAT CONTRAST** control, and LED **POWER** indicator are on the front panel. All connections to the printed circuit board are made with connector blocks, simplifying construction and allowing the board to be removed in a few minutes for troubleshooting or experimentation. (*photo courtesy of A & A Engineering*)

ing. Parts can be removed after soldering, but it is tedious and you may damage the board or the part during removal. The use of sockets simplifies matters for integrated circuits. If the IC chip is backward and power is applied, there is a high probability the chip will be destroyed. None of the IC chips in this project are very expensive, but it is very frustrating to wait for a $2 replacement chip when everything else is ready to go.

Soldering: Make good solder joints! What if you identify and orient the parts perfectly and the board still doesn't work? Is it a bad component? Possibly, but it isn't common. The real culprit is probably one or more bad solder joints! The board manufacturer has done everything possible to help you make the proper

joints. All pads are plated so the solder will flow easily over the pad and down into the hole. The green solder mask is there to assure that solder will not flow to adjacent pads where it doesn't belong. A good solder joint is smooth and shiny, with the solder wicked around the component lead and flowing on the surface of the pad and down into the hole. A dull, crystalline, or bumpy appearance means a poor joint and the possibility of a problem. Good solder joints, together with proper identification and orientation or parts, are your best guarantees of a successful project.

I suggest the following order for mounting components on the circuit board:

1) All connectors, including P1-P4, J1 and J2, and the IC sockets. Be careful to orient the sockets and

connectors as indicated on the silk-screened parts layout.

2) The two, multi-turn printed-circuit potentiometers (Ro and Rg) and all test points.

3) All fixed, ¼-W resistors.

4) The three aluminum electrolytic capacitors near P1. Note that these units are polarized and the + leads *must* go in the holes indicated,

5) All monolithic, mylar, disc, and dipped mica capacitors. These are non-polarized units so just make sure each goes with the proper set of mounting holes.

6) The silicon and germanium diodes. Note that these units are polarized and that the banded end of each should be aligned with the band on the silk-screen layout. These are the most delicate components of the project so handle them carefully and don't overheat them when soldering.

7) The trimmer capacitor (C25) and the crystal (Y1).

The kit comes complete with a drilled and punched, silk-screened cabinet and all necessary mounting hardware. Figure 5.5 shows the completed unit, including the small ac power supply. The cabinet measures 7 (W) × 5.5 (W) × 2¾ (H) inches (17.8 × 14.0 × 7.0 cm). The SAT and HF jacks and P4 are accessed at the rear apron. These components mount on the main board, so there is no interconnecting wiring. The power supply module mounts on one side of the cabinet and the ac power cable, fuse holder, and ON/OFF push-button switch mount on the rear apron, behind the power supply.

The AM/FM selector switch (SW1), SAT CONTRAST control, and the POWER LED mount on the front panel. Most of these components mate to the main board with a single connector (P3). All connections to the board from the accessory power supply are through a single connector. The unit can be assembled and disassembled in a few minutes. This is very handy for troubleshooting or experimental modifications.

Before installing any of the ICs in their sockets, verify the proper polarity and value for the various power supply voltages. Take special care to avoid shorting or accidental contact with ac wiring. The A&A power supply module is equipped with a flexible plastic shield to avoid such contact. AC line voltages and current *are potentially lethal*, so exercise care and judgment. All dc voltages should measure within ±10% of the nominal values with 5% being typical. Once supply voltages have been checked, turn off and unplug the unit from the ac mains. Insert the nine integrated circuits, making sure the notched or marked end of each device corresponds to the silk-screen layout.

IMPORTANT OPERATIONAL NOTES

Both check-out and operation require that the interface connect to the computer using a standard parallel cable with DB-25 male plugs at both ends.

Such a cable is supplied with wired-and-tested units from A&A and they are also universally available from computer outlets. The use of shielded cables will reduce the potential for RF interference from the computer, particularly for HF (short-wave) reception. Having an effective common ground for all equipment (see the *ARRL Handbook*) is also a big help.

If you want to operate a printer *and* the interface, there are three methods available. First, a parallel switching box can be used to connect one port to both devices. These switches are often called "A/B" switching units. Alternatively, you can purchase an inexpensive parallel printer port card for either unit. Finally, you can always swap connections to a single port, as needed. This is the least satisfactory approach and leads to premature wear on connectors if carried to extremes. The maximum length of parallel cable that will work reliably has not been determined. Six-foot cables work well and a pair of such cables in an A/B switch installation would probably be fine, although the use of shorter cables is probably a good idea.

When installing connectors, *both the computer and the interface should be off!* Once the cable has been installed, you may turn the units on in any order.

CHECK OUT

If you have purchased a wired-and-tested interface, the following check-out steps will not be required. You may wish to read through this section, however, in case you should ever wish to do any service work on the unit.

Clock Adjustment

The simplest way to adjust the clock oscillator is to use a precision frequency counter. By precision, I mean one whose reference oscillator has been precisely calibrated. Setting the interface clock circuit to what appears to be the proper frequency is a wasted effort if the counter is inaccurate. Setting the clock is a one-time adjustment, so it is entirely practical to bring your unit to a local radio amateur or electronics service shop to get access to a frequency counter.

Connect the input of the counter to TP3 and the adjacent TPG (ground) test point. Using an insulated tuning tool, *carefully* adjust the trimmer capacitor (C25) near the crystal to obtain a reading as close as possible to 4.194304 MHz. Some counters may be limited to a 6-digit display. In that case, set the counter for the kHz display and adjust for a reading of 194.304. Although it may be difficult to get the reading this precise, you should easily be able to adjust it to ±25 with respect to the last two digits. The closer you can get, the better.

It is also possible to set the clock while observing a satellite image display. Check out the video circuits in the section that follows, and then refer to Chapter 9 and make a VCR recording of a satellite pass. If you play this recording back into the unit and display the picture, the image will probably be tilted slightly (typically angling from right to left), reflecting the error in the absolute clock frequency. Play the recording through the unit again and *carefully* adjust C25 in small increments, striving for a vertical display. You will probably have to play the tape several times to get the clock precisely adjusted, but the tape makes it easier than waiting for live satellite passes.

If you have GOES, METEOSAT, or GMS receiving capability, these images come along very regularly, making it practical to adjust the clock using the signal output from your receiver. The apparent clock error, in the form of slanted images, varies with mode. WEFAX, requiring a display of about 200 seconds (3 minutes, 20 seconds) of image data, is the least demanding mode for clock stability. In contrast, TIROS or METEOR satellites require almost six minutes to display their images. A very slight tilt on your WEFAX display will be almost twice as obvious when using one of these modes, due to the longer display time. The most demanding mode of all, for the clock, is HF WEFAX display, since a full-screen image takes about 12 minutes to display!

From this point on, the easiest approach to finishing the setup of the interface is to use the oscilloscope display function in the WSHFAX program. Chapter 9 covers installation of the interface and the installation and booting of the software.

Parallel Port Test

The first thing the WSHFAX program will do is to attempt to locate the parallel port to which the interface is connected. There are three possible I/O addresses for parallel ports in an IBM-compatible computer. The program checks each one in turn, looking for the CLK signal on the port ERROR line. It will spend several seconds checking each port address. If the ERROR line does *not* change its logic level, a *no interface* notice will be posted for that port.

If WSHFAX finds a valid port, a verification notice will be posted (see below). If it fails to find a valid port when all three possible addresses have been checked, it will post a "PARALLEL PORT INTERCONNECT ERROR" message and list a number of possible reasons for the failure. Listed below are the most common problems, along with a troubleshooting guide.

No Power. If you failed to apply power to the interface, the test will fail. Turn the interface on and hit the **T** key to repeat the parallel port test routine.

No Parallel Cable. If you did not interconnect the interface and computer with a parallel cable, the test will fail. Power down, connect the cable, power up, and reboot WSHFAX.

Defective or Improper Cable. A defective cable is not likely, but it is remotely possible that you have used a cable designed for a *serial* port. These will not have all the interconnections used by a parallel port cable. Double-check your cable.

Clock Circuit Problems. If you have attended to the earlier adjustment of the clock frequency, you have verified the proper operation of the oscillator and buffer stages (U4). If you skipped this because of the lack of a frequency counter, use a logic probe to check the functions of U4, U5, and U7B.

Demodulator Tests

If your interface passes the parallel port test, you will be presented with the WSHFAX Main Menu options. Key **O** to invoke the digital oscilloscope feature. An oscilloscope-like display will be posted to test the demodulator and data I/O circuits. This display shows a time interval equivalent to one METEOR, NOAA, or HF WEFAX line or two GOES WEFAX lines. The bottom line of the display (black) is equivalent to 0 V while the upper line (white) represents +5 V.

AM (Satellite) Demodulator. Connect a test lead between TP4 near U5 and the center pin of the **SAT** input jack (J1) on the rear apron of the interface. Preset the **AM/FM** switch on the front panel to AM and the interface **SAT CONTRAST** to minimum (maximum counterclockwise). Slowly advance the **SAT CONTRAST** control while observing the computer display. You should see the indicated signal level begin to rise from the lower line (black) as the control is advanced, smoothly increasing to the top (white) limit. If you don't get an increasing signal level, connect a multimeter to TP2 and repeat the cycling of the **SAT CONTRAST** control from the minimum position. The indicated dc voltage should start near 0 V and rise smoothly to +5 V. If it does, your AM demodulator circuits are working and the problem is related to U3, U6, or the control signals produced by U7. If the dc voltage measurement does not behave as indicated, you will have to follow the input signal through the various AM stages.

FM (HF) Demodulator. Calibration of the FM demodulator circuits requires an accurate source of 1500 Hz (black) and 2300 Hz (white) audio tones. The easiest approach, if you own or have access to the test equipment, is to use an audio signal generator as the tone source. If the generator is one of the newer digital units, it can be set directly to the necessary frequency. An analog generator can be monitored with a frequency counter to assure accurate tone frequencies as well.

Alternatively, if you have a modern, synthesized short-wave receiver (you probably do if you are interested in HF WEFAX), the receiver can be used as a source of accurate audio tones. Connect a test lead to TP3 and place it next to a short wire connected to the receiver antenna lead. Set the receiver for USB (upper sideband) and adjust the tuning control for an indication of 4.194300. At this point, if the receiver is tuned to 4.192800 MHz, a loud 1500-Hz tone will be produced by the receiver. When the receiver is tuned to 4.192000 MHz, a 2300-Hz tone is heard.

1) With 1500 Hz input, adjust the multi-turn **OFFSET** pot (Ro) until the horizontal level indicator on the screen display rises *just* above the black baseline.

2) Now apply 2300 Hz input and adjust the multi-turn **GAIN** pot (Rg) to a point where the level indicator is *just* below the white (top) reference line.

Go back and forth between steps 1 and 2 several times since there is some interaction between the settings. You are done when a 1500 Hz input indicates just above the baseline of the display while 2300 Hz registers just below the top limit. If the tone source is varied between 1500 and 2300 Hz, the indicated level will shift smoothly between the black and white limits.

At this point the interface is complete and you are ready to display images. Refer to the operating instructions in Chapter 9 for the use of the various functions provided by the WSHFAX program.

WSH INTERFACE PARTS LIST

All the electronics components required for the interface can be obtained from the parts vendors listed in the Appendix. For the sake of convenience, parts numbers from the Digi-Key Corporation catalog have been included in the listing that follows.

The list does *not* include the cabinet and mounting hardware. Since the parts list is most useful to those scratch-building the interface or using just the PC board from A&A Engineering, connectors P1 through P3 are also omitted since these connections would be made with point-to-point wiring. If you elect to buy the complete kit offered by A&A Engineering, all parts and hardware are included.

In each parts category, the first entry is the required quantity, followed by the component description or value. The Digi-Key catalog number is the final entry. Prices for the listed items have not been included, since they are subject to change. At the time the list was compiled from the July/August 1993 Digi-Key catalog, all of the items came to a total of approximately $90.

Integrated Circuits

4	LM324N quad op amp	LM324N-ND
1	ADC0804 8-bit A/D converter	ADC0804LCN
2	74LS00 quad NAND gate	DM74LS00N-ND
1	74LS157 quad 2 to 1 multiplexer	DM74LS157N-ND
1	CD4020 14-stage binary counter	CD4020BCN

IC Sockets

6	14 pin DIP	AE8914
2	16 pin DIP	AE8916
1	20 pin DIP	AE8920

Diodes

4	1N270 germanium	1N270
2	1N4148 silicon switching	1N4148
2	1N5231B 5.1 V zener 500 mW	1N5231BPH
2	1N5237B 8.2 V zener 500 mW	1N5237BPH

1/4-W, 5% Fixed Resistors

1	100 Ω	100Q
11	1000 Ω	1.0KQ
1	1500 Ω	1.5KQ
5	2200 Ω	2.2KQ
2	4700 Ω	4.7KQ
1	6800 Ω	6.8KQ
17	10 kΩ	10KQ
1	12 kΩ	12KQ
2	15 kΩ	15KQ
2	22 kΩ	22KQ
1	27 kΩ	27KQ
2	39 kΩ	39KQ
2	100 kΩ	100KQ
1	1 MΩ	1.0MQ

Potentiometers

| 2 | 1-kΩ, 12-turn pc mount | 3266W-102-ND |
| 1 | 10 kΩ panel mount | RV4N103-ND |

Capacitors

1	180 pF (substitute for 180 SM)	P4803
1	470 pF ceramic	P4808
1	2.2-22 pF trimmer	SG3003
2	0.001 µF dipped mylar	P1000
3	0.0047 µF dipped mylar (sub. for 0.005M)	P1008
2	0.01 µF dipped mylar	P1012
1	0.015 µF dipped mylar	P1014
4	0.022 µF dipped mylar	P1016
2	0.047 µF dipped mylar	P1020
1	0.1 µF dipped mylar	P1024
2	0.22 µF dipped mylar	P1028
18	0.1 µF metallized film	P4525
4	1 µF non-polarized	P4537
2	47 µF alum. 16 V radial mount electrolytic	P6226
1	100 µF alum. 16 V radial mount electrolytic	P6227

Miscellaneous

1	SPDT miniature toggle switch	CKN1004-ND
2	Panel-mount phono jacks	SC1134-ND
1	4.194304 MHz microprocessor crystal	X007
1	DB-25F PC-mount connector	AN2103-ND

1	Green panel-mount LED	L10035-ND	
1	+5, +12, and −12 V ac power supply (Use TS9965-ND in 220 V ac countries)	TS9960-ND	
1	3AG Panel-mount fuse holder	F005-ND	
1	Pkg. (5) 3AG 0.25 A fuses	F109-ND	
1	3-wire grounded ac power cable	C106-ND	

A&A ENGINEERING PC BOARD AND KIT OPTIONS

You can purchase printed circuit boards, parts sub-kits, complete parts kits, or wired and tested units for the WSH Satellite Interface. Complete parts kits include the WSHFAX software package. As noted below, the WSHFAX software is available as a separate item for those scratch-building the interface or those who have purchased the PC board or parts sub-kits.

Part No.	Description	Price
200-PCB	Blank circuit board only	$19.95
133-PCB	Power supply circuit board	$7.95

200-KIT	Complete kit: all circuit boards, parts, cabinet and hardware (includes power supply) and the WSHFAX program.	$159.95
200-ASY	Complete wired and tested interface unit with WSHFAX software.	$189.95
169-KIT	Optional 20-segment LED tuning indicator ($40 when added to wired and tested interface option)	$35.00
WSHFAX	program (included with kits and wired and tested units)	$35.00

240 V version: add $5 to above prices

A&A Engineering
2521 West La Palma, Unit K
Anaheim, CA 92801
tel 714-952-2114
fax 714-952-3280

Chapter 6

Programming the WSH Interface

A complete software package is available for the WSH interface unit, so you can put it to use without any programming effort or experience. This chapter is not required reading. Instead, it serves three purposes:

1) To satisfy user curiosity about how the software functions.

2) To assist anyone who would like to develop their own custom software for the interface.

3) To serve as a guide for programmers of non-IBM-compatible computers in developing software for their systems.

This chapter is *not* intended to be a tutorial in either BASIC or assembly language programming. I assume you know the basics of the subject. My discussion centers on programming examples specific to the interface project. I should note, however, that one of the best ways to learn to program is to study programming guides and manuals.

Most programming guides have you writing happy faces on the screen and other arcane tasks. Few of these exercises are relevant to the problem of reading data from the interface and applying that information properly. The tasks outlined here can serve as very specific exercises to help you develop some useful programming skills. You can also build a library of essential routines needed to create a fully functional program.

The topics I address in this chapter are essential to a working program, but are not a complete program structure. The program listing for the interface software, WSHFAX.EXE, is equivalent to about three chapters of text in this edition. Listing the program, let alone explaining each step, would demand more space than we have available. Instead, I will concentrate on the tools required to accomplish certain essential tasks. Beyond that, it's up to you to determine exactly what you want your own program to do. In that respect, the information provided in Chapter 9

regarding the features of the WSHFAX program may prove interesting.

PROGRAM LANGUAGES

The programs relevant to satellite image programming are varied. The major players are BASIC, C, and Assembler. BASIC is familiar to most computer users and BASIC programs are useful for a number of tasks, including orbital prediction. BASIC comes in two major forms: *interpreted* and *compiled*. Interpreted BASIC works by having the program statements converted (interpreted) into the machine language of the microprocessor as each command, directive, or instruction is executed. This continuing translation process slows down execution, resulting in garden-variety BASIC being one of the slowest languages. While more than adequate for orbital predictions, interpreted BASIC is far too slow to actually handle image data.

Compiled BASIC is a different animal. The language looks quite similar to other forms of BASIC. Once the program is written, however, it runs through two other programs: a compiler and a linker. Together they create the machine-language equivalent of each statement in the program in a single operation. No conversion is required when you finally execute the program.

Compiled BASIC is much faster than the interpreted variety, which expands the ways we can use it. It is still too slow for real-time image manipulation, though. Microsoft *QuickBASIC* is an example of compiled BASIC with a very powerful programming, editing, and debugging environment. I use this program extensively for doing menus, disk I/O, and other functions that are not as speed-sensitive.

Assembly Language Programming or simply "Assembler" is one step removed from the machine language the microprocessor actually uses. The result is a very compact program code when the final program runs. Assembly routines run *very* fast and are more than adequate to keep up with real-time satellite video—even on a slow microprocessor such as the 8088 or 8086. Within a family, such as the Intel 80XXX, new instructions are added for more powerful processors. Assemblers geared for the 80386 and 80486 have instructions that are not covered by earlier processors such as the 8086/8088. Although the new extensions take full advantage of the newer processors, they can cause problems if the program executes on an older system.

I do all my video-handling routines in the 8088/8086 version of Microsoft's *Macro Assembler* (MASM). I can always get the needed speed without accessing advanced instructions that older processors cannot use. It's very tedious to write a complex program in assembler. Instead, I compile a *QuickBASIC* module for most aspects of the program, linking assembly modules (which have the speed to handle live satellite video). The result is a single program that combines the convenience of BASIC and the speed of assembler programming. It isn't too difficult to write assembly modules to handle most aspects of the interface. The reward comes when you see how easily you can tap the speed potential of even a tired 8088 PC!

There are differences within a family of processors when it comes to assembly language features. This is trivial, however, compared to major differences in most aspects of the languages used for different microprocessor families. For example, an assembler routine suitable for the entire range of Intel processors is often useless for the Motorola 68XXX family. That is one of the reasons why so much of the software for IBM-PC compatibles is not replicated for Mac and Amiga systems (which use various Motorola CPUs). Even so, learning assembler for any microprocessor reduces the time required to learn a new version for another family.

C (in all its variants) is very popular with professional programmers. It combines the relative ease of a language like BASIC (with pseudo-English commands and statements) with the speed of assembler. C is a compiled language, but, unlike BASIC, the resulting code can replace many assembly modules in all but the most time-critical functions. C modules can be linked with small assembler modules where very-high speed is required. Modules in either language can be linked to other compiled BASIC modules if you have the right software. Besides speed, the major advantage of C is *portability*. A properly written C source code module often runs on widely different microprocessors if each has a compatible C compiler. The wide usage of C permits many common applications to run in a very similar fashion on computers such as the PC and Mac. I will not present any C code in this chapter, but experienced C programmers will have no trouble writing equivalent routines for much of what will be presented.

UNDERSTANDING THE PARALLEL PORT

In creating an I/O port to drive accessory parallel printers, the designers of the original PC had to deal with 8 bit data ports. The printer data, status, and control formats were already well-defined within the printer industry, so the computer designers' task was to meet those requirements. To do so, they created a *Parallel Printer Port* standard that is actually three ports, from the perspective of the computer's CPU.

This three-in-one port is located at one of three addresses within the I/O address space of the original 8088 CPU:

Hex	Decimal
03BC	956
0378	888
0278	632

During the *power on self-test* (POST), the system checks each of these possible addresses, looking for parallel port hardware. The first port address where it encounters the appropriate hardware is designated LPT1, the second LPT2, and so on. Since there are only three such addresses, there can only be three parallel printer ports installed on any given system. Such ports were "add-ons" (via interface cards) on the original PC. Today there is likely to be one port implemented via hardware on the motherboard. On an IBM system this will probably be 03CBh. This may not be the case with specific clones, many of which place their motherboard port at 0278h or 0378h.

There are a number of different ways to tell what port addresses you have installed in your system. The really important question is, which one has the interface connected and working? In just a bit, I'll show you how to locate the interface port.

The three addresses noted above are really base addresses. Each marks the start of a sequence of three ports that make up the parallel printer "port." These three ports have the following function and structure relative to the base address:

ADDRESS	base	base+1	base+2
FUNCTION	Printer Data	Printer Status	Printer Control
MODE	output	input	output
BIT 7	D7	busy	not used
BIT 6	D6	acknowledge	not used
BIT 5	D5	paper out	not used
BIT 4	D4	printer ready	enable irq
BIT 3	D3	error	select
BIT 2	D2	not used	initialize
BIT 1	D1	not used	auto line feed
BIT 0	D0	not used	strobe

The base address port is normally dedicated to the output data to the printer. In computers from the AT onward, it can be configured as an input port as well. Since this capability can't be used by all IBM systems, I elected to ignore the base port.

The port at base + 1 consists of 5 input lines normally used to read the printer status. The interface uses bit 3 of this port for the clock signal (CLK). Bits 4-7 are used to input 4 bits of video data from the multiplexer that follows the A/D converter.

The base + 2 port consists of 5 output lines. We will only use 2 of them. Bit 0 serves to strobe the interface A/D and bit 1 is used to toggle the multiplexer output to select the video nibble (4 bits) to be read.

Normal operation of the software involves reading up to 5 bits of data from the interface at base + 1 and controlling the interface with 2 bits of output data at base + 2. This applies to the IBM-compatible parallel printer port. Since we only require a total of 7 bits of data (5 input and 2 output), connecting another kind of computer to the interface can be accomplished with only 7 programmable I/O lines. A section at the end of this chapter is devoted to illustrating how best to accomplish connection to other computers. For the moment, let's look at the IBM-compatible software requirements for communicating with the interface.

PROGRAM STRUCTURE

As noted earlier, my own software always consists of at least two modules. The main body of the program is written in compiled QuickBASIC. This module handles all the menus, keyboard and disk I/O, and almost anything else that doesn't require the warp-speed achieved with assembly language programming. The second module is written in assembly language. This module handles the real-time acquisition of image data, screen dumps and other operations where speed is important. The sections that follow illustrate fundamental operations performed by both pieces of the program, beginning with the BASIC program module. Any of the functions accomplished in the BASIC module could be handled just as well in C. Also, there is no reason why the entire program could not be written in assembly language.

In the sections that follow, I provide numerous examples of code fragments that accomplish certain basic tasks. In many cases, the code fragment is small enough to fit into the main text. Where the listing is too long, it is included in a block of program listings at the back of the chapter. These listings cover both BASIC and assembler (ASM) code samples and are thoroughly commented.

ESSENTIAL FUNCTIONS
OF THE BASIC MODULE

Microsoft's QuickBASIC is a thoroughly modern version of BASIC that, among other things, allows you to dispense with line numbers. It uses symbolic labels for jumps (GOTO) and subroutines (GOSUB). This makes for much cleaner source code. The code samples I offer follow this convention. It is the BASIC

module that provides all the look and feel of the final product and you will probably do a lot of tinkering with it. Remember, my goal is not to show you how to write a complete program, but simply how to perform the essential functions associated with the interface and the assembly language module.

Defining the Assembly Language Module

The BASIC compiler needs to know that there is going to be an assembly language module. You must tell it at the very beginning of the BASIC source code. Assuming your assembly source-code module will be called "sat," these two lines of code will get the job done:

```
DEFINT A
DECLARE FUNCTION sat (a%)
```

Of course, your code block can have any label you wish, but the names in the BASIC and assembly modules *must* agree. Note that the name in question is NOT the name of the assembler source or object file. It's the label you provide for the code segment in the assembler source listing. If the names do not agree, you will get an "unresolved externals" error message when you attempt to link the two modules.

Finding the Interface

If you know the address at which your parallel port is implemented, you might think that you could simply code that into your listing:

```
port = &H378
```

There are two—no, make that *three*—problems with this approach. First, your documentation may not tell you the address. You'd have to use DEBUG or some other means to find the address. Second, hard-coding of the port address won't allow you to switch between computers if the other machine uses another port. Finally, and most important, knowing the port address does not mean that the interface is properly connected. It also does not tell you that its power is on and that the clock is working. If you jump to the assembly language module without a working clock, the computer can "hang up" and require a keyboard reset or even a complete power-down and reboot.

A solution to this problem can be found in listing BASIC 1, a port testing module. The module accomplishes the following:

1) The variable <port> is initially assigned a dummy value of zero.
2) All three possible port addresses are checked in sequence for the presence of the interface:
   a. With up to 100 attempts, does bit 3 of the status port show a HIGH?

b. With up to 100 attempts, does bit 3 of the status port show a LOW?

3) If both (a) and (b) are satisfied for a specific base address, the interface has been found and is operating. The variable <port> is assigned this address value and a jump is executed to the next stage of the program.

4) If no port passes the test, the interface is not installed, the cable is defective, power is not applied, or the clock isn't working. In that case, <port> retains the value of 0 and you jump to an error message.

Here is where you can get creative, if you wish. The error message can list all the possible problems and invite a retest (with a jump back to the port test module). Or you can continue (to permit the use of other aspects of the program), or simply quit. In any case, this routine will locate the interface without any special configuration files. It runs on different machines with any of the three possible port assignments.

Interface Set-up

Operation of the assembly language module can be much easier if we could assume a certain set-up configuration for the interface. The two conditions listed below are most desirable:

1) Strobe input HIGH to place the A/D converter in standby
2) Nibble Select LOW to read high nibble data by default

Assuming the port test has been passed, the following code fragment provides the interface setup:

```
configure:
    OUT (port + 2), 0
```

Port + 2 is the control output port, while sending a 0 to the port will set the bits properly. Why this is so is a bit complex, but we will deal with it later.

Setting the VGA Display Mode

We can set this display mode from within the assembly language module, but this is a *very* complex procedure. In contrast, the following BASIC fragment can accomplish the job easily:

```
screenset:
    SCREEN 12
    CLS
    r = 0
    FOR n = 0 TO 60 STEP 4
    c = INT((65536 * n) + (256 * n) + n)
    PALETTE r,c
    r = r + 1
    NEXT n
```

Screen 12 specifies the 640×480 display mode. The rest of the routine sets the palette so pixel values between 0 and 15 will produce a linear grayscale from black to white.

Video Look-up Tables

The interface produces video values ranging from 0 (black) to 255 (white). The screen display can only handle values from 0 to 15, so we must convert from A/D values (0-255) to screen display values (0-15). This conversion provides a tremendous range of power, since the conversion technique greatly alters the image display format.

The fastest way to accomplish the conversion is to have a table of values that correspond to the desired display values. If the A/D provides a value of 133, for example, the program checks entry 133 in the table to retrieve the screen display value. As we will discuss in Chapter 10, the relationship between input and output values can be anything you need to achieve a specific result. Table data can be derived mathematically or manually. You can compute table values "on the fly," or custom table data can be stored and retrieved from disk. Whenever you first display an image from the receiver, it is useful to use a linear relationship between the A/D values from the interface and screen display values. A linear conversion table can be "poked" into memory as follows:

```
default_table:
    DEF SEG = &H5000
    memptr = &H100
    FOR n = 0 TO 255
        v = INT(n / 16)
        POKE (memptr + n), v
    NEXT n
    RETURN
```

The table resides in segment 5000h and the start is offset by 100h (memptr). This avoids conflict with the mode and port values, which we'll insert into the same segment in the next section. If this subroutine is called before any of the active imaging modes, the new picture data will be displayed in a linear fashion. As we will see in later sections, while the system displays 16 grayscale values per pixel, the original 256 grayscale values are stored in a RAM buffer. If we do a screen dump from the buffer, having poked another table into RAM, the result is an image processing function (a subject discussed in greater detail in Chapter 10).

Getting to the Assembly Language Routine

Functions such as menu displays, keyboard input, and disk I/O can all be handled in the structure of the BASIC program. There is a whole list of functions, however, that must be performed within the assembly language module:

1) Input as display of image data in a variety of modes
2) "Dumps" from the RAM buffer to the display screen
3) "Oscilloscope" displays
4) Image inversion or rotation
5) Performing various image processing functions

Items 1 and 3 need the speed of assembly language to even function at all. The rest of the functions *could* be performed in BASIC, but they would take too long. A RAM-to-screen dump might take 6 minutes in BASIC as opposed to 10 seconds in assembler. Although you could construct an assembly language module for each possible function, this would represent a very inefficient approach to programming. The easiest approach is to add functions to a single assembly module.

Two items must be passed from BASIC to the assembly language module: a numerical value to indicate what function needs to be performed, and the base address of the parallel port. The scheme I suggest uses the variable <mode> to signify the function to be performed. For reasons that will be evident shortly, I allocate mode values from 0 to 10 for functions that don't use the clock-screen dumps and the like. Values above 10 are used for any routines that require the clock.

An elegant programming technique passes mode and port data between the two modules using the stack. The only trouble with this technique is that it is fairly easy to mess things up when you are just getting started! An approach that works a bit more reliably, at least at the start, is to put the mode and port data at a specific spot in memory. Follow that with a jump to the assembly module. One of the first things the assembly language program does do is retrieve the mode and port data so it knows what it's supposed to do. Listing BASIC 2 shows how these functions can be accomplished.

The first step is to check both the mode and port values. If the value for <port> is 0 (no clock) and the value for <mode> is greater than 10 (the clock is required), the program jumps back to the Main Menu. This avoids entering the assembly module and then hanging up because no clock is present.

The second step is to break the <port> word (two bytes) into individual bytes. The program then proceeds to set the pointers for the mode and port data. In the example shown, the segment value is 5000h and the RAM pointer within the segment is 0. The program then pokes the mode value to 0, the low-order port byte to 1, and the high-order byte to 2. With this accomplished, the routine jumps to the assembly module.

When a return is executed, the all BASIC variable values are still intact. The program can branch to various points, depending on the <mode> value.

Testing

These are the most essential functions, leaving a full-range of work for creating a complete program structure. Even at its early stages, the BASIC module is fairly complex. Obviously, you would like to test and debug the BASIC module prior to dealing with the extra complexity of the whole package!

If you try running the BASIC module on its own, you will get error messages because the assembly module is not present. To avoid this, edit the following program lines as indicated:

REM DECLARE FUNCTION sat (a%)
.
.
.
REM x = sat (a)

Inserting the REM statements disable the DECLARE FUNCTION directive and the entry to the assembly language module. You can now test and debug the BASIC module with ease. When the BASIC code is working smoothly, remove the REM statements prior to compiling the module.

ESSENTIAL FUNCTIONS OF THE ASSEMBLY LANGUAGE MODULE

While the BASIC program module defines how the program interacts with the user, the assembly language module determines how well the program actually functions. The complexity of the module is determined by how cleverly you craft the subroutines. A subroutine that writes video values to the screen, for example, could also increment the pixel count and line count (if required). That way, such housekeeping tasks become automatic and you don't need to perform the functions in another routine.

As in the previous section, the goal here is *not* to teach you to program in assembly language, or how to construct a complete module. What I do provide is a series of reliable routines to accomplish the most basic tasks that will certainly need to be performed.

The Proper Overhead

There is nothing more literal than an assembly language source listing. You can cause problems right away if you don't handle certain overhead problems right at the beginning of your code. Listing ASM 1 shows how to start a typical source listing for a single module linked to a BASIC program. This listing accomplishes the following:

1) It declares the model type, using the new, simplified memory models implemented by Microsoft.

2) Defines the stack area for the assembly module.

3) Defines the labels and type (db = byte, dw = word) for all data elements. Note that the ones shown here are only those used in the sample listings.

4) Labels the start of the CODE segment. Note that the name used here must match your DECLARE FUNCTION directive in BASIC.

Initializing the Assembly Program Module

Having taken care of the necessary overhead to assure the validity of your assembler source code, it is time to get down to business. Listing ASM 2 shows one approach to getting the program on the road. This assumes you have cached mode and port data as suggested in the previous discussion of the BASIC program module.

The first step is to fetch the mode data (which defines what we want the module to do) and the port base address from segment 5000h, where they were deposited by the BASIC module. The mode value is stored in *mode*. The port base address is incremented by 1 and stored in *status*. It is then incremented once more, with the value stored in *contrl*. Now, if we need to input data from the STATUS port, we fetch the address from *status*. We can do precisely the same thing when we need to output data to the CONTROL port.

The rest of ASM 2 is some basic initialization so that we don't have to remember to do it elsewhere in the module. I use segment 6000h (*bufseg*) as the start of my video buffer in RAM, beginning at location 0 in that segment (*bufptr*). The initialization routine sets these starting values.

Finally, the listing initializes the screen values. I display data in a 512×480 window on the far right of the 640×480 VGA screen. The starting line is 0 (set in *lcount*) and the starting pixel is 128 (set in *pcount*).

At this point, the *mode* value is retrieved. From here, I run through a list of comparisons. *Scan* jumps to different routines in the assembly module, depending on the mode value. The last item, *scanx*, executes a RETURN instruction that takes us back to the BASIC module. If the value in *mode* doesn't match any of the choices in the assembly module, you will be bounced back to the BASIC module. This isn't likely, but it eliminates the possibility of implementing some routine that you *don't* want! Whenever you're ready to return after completing a task, executing a jump instruction to *scanx* will accomplish an orderly exit from the module.

Keeping Track of Time

Precision timing is all-important in the assembly language module. All operations, such as sampling pixels, are performed by referencing the interface

CLK signal on bit 3 of the STATUS port. Listings ASM 3, ASM 4, and ASM 5 are all time-related functions.

The most important timing operation is keeping track of the status of the CLK signal. ASM 3 is a subroutine (*chi*) that determines when the CLK line is HIGH. ASM 4 is an almost-identical function (*clo*) to determine when the clock is LOW.

We begin by pushing all the registers on the stack. This allows us to use the registers for one thing in a main program routine, while using them for something else in a subroutine. At this point, the address of the STATUS port is loaded into *dx* and the *al* register is loaded from the STATUS port.

Only bit 3 is the clock data, so an AND instruction is used to mask out all other bits. If the CLK line is high, *al* contains an 8. If it is LOW, *al* is 0. If the subroutine finds an appropriate value, it exits. If not, it keeps checking the port until the proper CLK level is noted. On exit, the register values are popped off the stack (in *reverse* order from the push operation at the start).

Listing ASM 5 uses both the *chi* and *clo* subroutines to implement precision time delays. On entry, register *dx* should contain the desired delay count (in CLK cycles). Let's say we wish to delay for one WEFAX line—1024 clock cycles. The time subroutine would call *chi* and *clo* and then decrement the count in *dx*. When *dx* reaches zero, we exit the subroutine. *Time* can be used anywhere in the program where you need a controlled delay period. If you only need to delay 1 clock pulse, it is usually easier to simply call *chi* and *clo* directly. For anything longer, simply load *dx* with the desired delay value and call *time*.

Sampling Video Values

We only need two types of data from the interface—clock status and video values. The *chi* and *clo* routines indicate how to read clock data. Listing ASM 6, the *getvid* sub-routine, shows the more complicated function of reading an 8-bit video value from the interface. It is more complicated because we need to read the port *twice*, once to get the high nibble (high 4 bits of the byte) and the second time to get the low nibble (the lower 4 bits).

After pushing the affected registers, we use *chi* to wait for the clock to go high. We then output to the CONTROL port to set the strobe line LOW. Here is where we run into one of the complications of working with the parallel port. If you refer back to Figure 5.3, you will see that the connection between the port STROBE line and the A/D WR control is through two NAND buffers (U7C and U7D). This should mean that to put a LOW on WR we would need to set the STROBE bit of the port LOW. Instead, you will see that I have set STROBE (bit 0) HIGH! This is because the parallel port *hardware* inverts whatever we write to bit 0. Thus, if we want strobe to be LOW, we must write a HIGH and *vice versa*.

Once STROBE as been set LOW, the routine executes a meaningless delay of 10 loops via decrementing the *al* register. This is to allow the data on the STROBE/WR line to settle. On today's very fast computers, the computer can move on before the comparatively fast TTL logic has time to fully stabilize to the new logic state. This 10-cycle loop will not overlay delay a slow machine but provides a settling interval for a fast one!

The STROBE line is then brought HIGH again by writing a LOW to control port bit 0 (remember the hardware inversion of the STROBE bit). This will cause the interface A/D chip to start a conversion. At this point we use *clo* to wait for the clock to go LOW. The reason is that the A/D chip requires between 70 and 100 microseconds to do a conversion. CLK is HIGH for 122 microseconds, so if we wait for a LOW on the clock, we know the output data from the chip will be valid.

Once the CLK line is LOW, we can proceed to input and format the video. It would be nice if we could read the video value in a single pass, but we are short available input lines, leading to the use of U6 (Figure 5.3) that requires that we read the 8 bit data in two passes — one for the high nibble (upper 4 bits) of the A/D data and the second to read the low nibble (lower 4 bits). U6-1 must be LOW to read the high nibble and HIGH to read the LOW nibble:

AUTO-FEED WRITE TO

READ	U6-1	OUTPUT	CONTROL
high nibble	LOW	HIGH	LOW
low nibble	HIGH	LOW	HIGH

Since U6-1 is connected to bit 1 of the control port (AUTO-FEED) through a single NAND buffer (U7A), that inverts the logic level applied at its input, we should simply have to set AUTO-FEED HIGH to read the high nibble and LOW to read the low nibble — easy to remember! Unfortunately, AUTO-FEED, like STROBE, is inverted in the parallel port hardware so we need to write a LOW to set AUTO-FEED HIGH and *vice versa*! Back in our BASIC module, we initialized control by writing a LOW to bit 1 (AUTO-FEED), so it should be HIGH, allowing us to read the high nibble by default.

The routine reads the byte from the STATUS port, the upper 4 bits of which represent the high nibble from the A/D converter. An AND 240 masks out the lower 4 bits, which include the clock bit (bit 3 of the byte) plus the lowest three bits which are not defined.

At this point we encounter another of the hardware inversion problems of the parallel port. Bit 7 of the byte is inverted in hardware—if it reads HIGH it is really LOW and *vice versa*! A few extra steps are thus required to invert bit 7 in software, prior to saving the data in *atod*.

Next, the routine toggle the AUTO-FEED bit to select the LOW nibble, inserting a delay loop to allow the logic data to settle. The read operation of the STATUS port is then repeated, complete with inverting bit 7. This bit, however, is the low nibble of the A/D data, so we execute four shift operations to slide the bits into the low-nibble of the byte. Data from *atod* is then added to al to complete the formatting of the 8 bits of video data and the byte is stored in *atod*.

Finally, AUTO-FEED is reset to the default condition so we don't have to remember the state we left it in, at which point the registers are recovered and the program exits the routine. GETVID is likely to be one of the more complex routines in any program you put together, in part because we need to read the STATUS port twice and format the video byte. The inversion problems with bit 7 of the STATUS port and bits 0 and 1 of the CONTROL port all add their little complications. If we take care of everything in the routine however, we can ignore the problems elsewhere in the program.

Converting 8-bit Video Values for 4-bit Display

Our A/D converter in the interface produces 8-bit video values (0-255), yet our standard VGA display can only handle 4-bit (0-15) data. Listing ASM 7, the *lookup* subroutine, uses our lookup table in RAM to make the needed conversion.

Recall from our earlier discussion about the BASIC module that the table was poked into segment 5000h, beginning at 100h and running to 100h + 255. In *lookup* we set the data segment up for 5000h and set the memory pointer (*bx*) to 100h—the start of the table. We then fetch the 8-bit video value from *atod* and *add* it to *bx*. If the video data were 0, we would be pointing to the start of the table in RAM. If it were 255, we would be pointing at the *last* table entry. We then fetch the table value from RAM into *al* and then store the result in *atod*. This value is in the form of 4 bit data, suitable for passing to the display. The relationship between the 8-bit input value and the 4-bit output value depends on how you constructed the data table, a subject to be discussed in greater detail in Chapter 10.

Writing a 4-bit Pixel to the Display

Writing the pixel value directly to the VGA display is a complex business in the 640 × 480 mode. This is the fastest way to do the job, but requires about 2-3 pages of code. We can perform the job easily, at a modest speed penalty, by using the pixel write routine in the computer's ROM BIOS. Listing ASM 8 shows how this is done, using a subroutine called *wrpix*. The routine we invoke is one of the BIOS *interrupt* routines—interrupt 10h to be precise.

We start by saving all the registers, for BIOS routines will use them in unpredictable ways. Once that is accomplished, we proceed to satisfy the requirements for the *write pixel* option for interrupt 10h:

1) Register *dx* must contain the screen line—we get that from *lcount*.

2) Register *cx* must contain the active pixel on the line—we get that from *pcount*.

3) Registers *ah* and *bh* must contain values of 0Ch and 0 respectively. These values are loaded directly to the registers.

4) The 4-bit value we want to write to the screen must be in *al*—we fetch the value from *atod*.

Now we can invoke the interrupt (*int*) and the BIOS code will write the pixel for us. We could exit at this point (after recovering the registers), but we can save ourselves a lot of trouble elsewhere in the program by updating pixel and line count values before we get out.

Reading and Writing to the RAM Buffer

During image reception, image data is written to the screen (4 bits) and to the RAM buffer (8 bits). Later, we might wish to read data from the buffer, in a sequential fashion, to process the image for display. Listing ASM 9 contains a subroutine to read data from the buffer (*rdram*) while ASM 10 documents the write-to-RAM (*wrram*) subroutine. Both are quite similar. If you have followed the routines so far, they should be pretty straightforward.

The current buffer segment (*bufseg*) and RAM pointer (*bufptr*) are retrieved to set up the read or write operation. Once data has been sent or fetched from RAM, the pointers are updated. If the RAM pointer (*bufptr*) is FFFFh, the segment is full. In that case, the segment pointer (*bufseg*) is updated by 1000h and the RAM pointer (*bufptr*) resets to 0. If the segment is not full, the value in *bufseg* is not altered and *bufptr* increments to be ready for the next read or write operation. Finishing this overhead work right after the read or write makes the rest of the program code easier to create. The pointer and segment management are embedded in the routines themselves.

A RAM-Dump—Putting It All Together

Writing all these subroutines might seem tedious, but having them on tap simplifies writing some of the routines in the main program code. For example, in a typical image processing routine, the BASIC module will poke a specific video conversion table into memory. At this point you ask the assembly module to dump the RAM buffer to the screen using the new table.

Listing ASM 11 shows a routine (*dump*) to accomplish this. This routine is very simple, precisely because we have a good library of subroutines to perform the various subtasks.

The routine calls *rdram* to get the 8-bit value from the RAM buffer. All the pointers increment to the next RAM location with this operation. The routine then calls *lookup* to convert the 8-bit value to a 4-bit value. Then *wrpix* is called to write the 4-bit value to the screen. Pixel and line pointers are updated automatically by this operation. Now the routine checks *lcount* to get the line count. If it is *not* 480, it will repeat the operation. If *lcount* is 480, the dump is complete and the routine executes a jump to *scanx* to exit the assembly module back to the BASIC program. Depending on the speed of your computer, this dump will take from 5 to about 20 seconds, as opposed to 5-15 *minutes* if we performed the same operation from BASIC!

Sampling a Pixel

During image acquisition, pixels must be sampled and passed to the RAM buffer as well as the screen. Listing ASM 12 documents *pixel*, another example of the power of linking subroutines.

The routine calls *getvid* to actually sample the pixel from the interface. It then calls *wrram* to write the 8-bit pixel to the buffer. Again, *wrram* takes care of managing the segment and RAM pointers. *Lookup* is then called to convert the 8-bit value to a 4-bit value. This is followed by *wrpix* to write the 4-bit value to the screen.

Take care to make sure that *wrram* precedes both lookup and *wrpix*. All 8-bit operations (like *wrram*) must be performed prior to using *lookup*, which will convert it to a 4-bit value.

Sampling a Line

As a final example of subroutine linking, listing ASM13 documents *wline*. This is a subroutine used to sample, store and display a WEFAX line.

Our clock operates at a frequency of 4096 Hz, which would let us obtain 1024 samples in the course of a 250 ms WEFAX line. If we take the simple case of a standard VGA display, a basic sampling strategy might be to sample a pixel and then delay 1 clock cycle. If we do this 512 times, we will display 512 pixels—a complete line—during a line interval of precisely 1024 clock cycles. In the listing for *wline*, *cx* holds the sample count. We call *pixel*, to sample the video (and also to store and display it), and then delay a clock cycle with calls to *chi* and *clo*. Now we decrement *cx*. If the result is *not* zero, the line isn't complete. We must repeat the sequence. If *cx* has counted down to 0, the line *is* complete.

At that point, we have another sampling decision to make. A WEFAX image will have 800 lines, yet we only have 480 screen display lines. One approach that will allow everything to fit, is to sample a line (as we have just done), and then delay for precisely 1 line. This results in the WEFAX frame using 400 of the available 480 lines. Our listing for *wline* takes this approach, so the active line is followed by a 1024 clock delay, implemented by calling *time*. With that, the line is finished and a return is executed. If we were to call this routine 400 times, we would display an entire WEFAX image.

Line routines for other modes are just as simple. If you insert checks of the value in *mode* at various points, the sampling strategy and time delay can be altered. Depending on the value in *mode*, you can use a single *line* routine that will work for all modes!

Of course, if you are programming for a more advanced VGA display, you have other options. A typical SVGA card permits a $1024 \times 768 \times 16$ display. You would still be displaying 4-bit pixels, but you could display *every* pixel on a WEFAX line and 768 lines. If you were willing to pass on the last 32 lines, you wouldn't need to skip any lines at all! Some of the new Hi-color VGA cards will let you display 256 "colors" at 1024×768, so you could directly display the 8-bit data. Once you know the basic principles of programming and have a collection of reliable software routines, you can accomplish almost anything!

Summary

While assembly programming is a new experience for most, it is really quite a bit of fun. There is nothing more satisfying than seeing code operate just the way you intended! With a collection of good subroutines, fairly complex programs can be built very easily. You might start out with a simple goal—WEFAX display. Once that is running, it takes much less effort to add new modes, and the program grows in power.

If you have some assembler programming experience, these routines outline all the essentials of communicating with the interface. Programmers for other computer families (like the Mac and Amiga) will find the interface just as easy to use with their systems. The last step is providing the connection between their computer and the interface.

INTERFACING TO OTHER COMPUTERS

If your computer has an IBM-compatible parallel port—or if you can add one—you have solved your interface problem. A parallel printer port that is fully compatible with the parallel drive requirements will almost certainly pass the compatibility test for use with the interface.

If you can implement such a port, you can implement an interface with a hardware card. Most cards allow you to implement a programmable 8-bit parallel port. Cards that permit operation of up to 24 bidirectional TTL lines (three 8-bit ports) are widely available. Although you can arrange the 8-bits any way you wish, I would suggest the following to maintain compatibility with most of the routines I have documented:

Bit 7	input	video (msb)
Bit 6	input	video
Bit 5	input	video
Bit 4	input	video (lsb)
Bit 3	input	clock
Bit 2	output	not used
Bit 1	output	nibble select
Bit 0	output	a/d strobe

The assembly language details will obviously differ, but the principles embodied in each routine are perfectly applicable.

By using interface options available for your own system, you can communicate with the interface in the same basic fashion as IBM-compatible systems. Most of the software techniques illustrated in this chapter can be implemented with little change. This speeds the software development process. Those with some programming experience on the Mac or Amiga will probably make themselves *very* popular. There is remarkably little satellite hardware/software available for these otherwise fine systems!

```
;================================================================
;      SAMPLE LISTING: BASIC 1
;            CATEGORY: main program listing
;                NAME: portcheck
;            FUNCTION: locates which port has the interface
;================================================================

portcheck:
            port = 0
            'TEST PORT ADDRESS 1
            tport = &H3CB
            GOSUB portest
            IF port <> 0 THEN GOTO configure
            'TEST PORT ADDRESS 2
            tport = &H378
            GOSUB portest
            IF port <> 0 THEN GOTO configure
            'TEST PORT ADDRESS 3
            tport = &H278
            GOSUB portest
            IF port <> 0 THEN GOTO configure
            'ALL ADDRESSES FAILED - POST ERROR MESSAGE
            GOTO porterror
portest:
            n = 0
            'MUST FIND A HIGH ON CLOCK/ERROR LINE TO PASS
portest1:
            v = INP(tport + 1)
            GOSUB bitchk
            if v >= 8 THEN GOTO portest2
            n = n + 1
            IF n < 100 THEN GOTO portest1
            'TEST FAILS - PORT STAYS 0
            GOTO portestx
            n = 0
            'MUST FIND A LOW ON CLOCK/ERROR LINE TO PASS
portest2:
            v = INP(tport + 1)
            GOSUB bitchk
            if v < 8 THEN GOTO portest3
            n = n + 1
            IF n < 100 THEN GOTO portest2
            'TEST FAILS - PORT STAYS 0
            GOTO portestx
portest3:
            port = tport
portestx:
            RETURN
bitchk:
            IF v >= 128 THEN v = v - 128
            IF v >= 64 THEN v = v - 64
            IF v >= 32 THEN v = v - 32
            IF v >= 16 THEN v = v - 16
            RETURN
```

```
;================================================================
;    SAMPLE LISTING: BASIC 2
;          CATEGORY: main program listing
;              NAME: mlentry
;          FUNCTION: sets up jump to assembly module
;================================================================

mlentry:
                'TEST IF THE REQUEST IS SAFE WITHOUT A CLOCK
                IF (port = 0) AND (mode >10) THEN GOTO mainmenu

                'BREAK PORT INTO 8 BIT VALUES
                portmsb = int(port/256)
                portlsb = port - (portmsb * 256)

                'SET CACHE MEMORY SEGMENT
                def seg = &H5000

                'SET POINTER TO START OF CACHE MEMORY
                memptr = 0

                'POKE MODE VALUE
                POKE (memptr), mode

                'POKE PORT VALUE
                POKE (memptr + 1), portlsb
                POKE (memptr + 2), portmsb

                'RESTORE DEFAULT DATA SEGMENT
                def seg

                'EXECUTE ML ROUTINE
                x = sat (a)

                'ADD MODE SPECIFIC BRANCHES IF NEEDED ON RETURN
```

```
;===============================================================
;     SAMPLE LISTING: ASM 1
;           CATEGORY: main program listing
;               NAME: none
;           FUNCTION: provides stack and data definitions
;              NOTES: must be at start of source code listing
;===============================================================

        .MODEL   MEDIUM, BASIC

        .STACK   100h              ;allocates stack

        .DATA

mode    db       ?                 ;mode value storage
atod    db       ?                 ;video value storage

pcount  dw       ?                 ;pixel count storage
lcount  dw       ?                 ;line count storage
status  dw       ?                 ;status (input) port address
contrl  dw       ?                 ;control (output) port address
bufseg  dw       ?                 ;default video buffer segment
bufptr  dw       ?                 ;default video buffer pointer

        .CODE

sat     PROC

                                   ;code listing starts here
```

```
;============================================================
;       SAMPLE LISTING: ASM 2
;               CATEGORY: main code listing
;                   NAME: setup
;               FUNCTION: recover and save mode and port data and set
;                         default values for the ram buffer segment,
;                         the ram buffer pointer, the starting screen
;                         pixel count, and the starting screen line
;                         count
;REGISTERS (ENTRY): doesn't matter
; REGISTERS (EXIT): all registers modified
;        SUBROUTINES: none
;            MEMORY: mode (db), status (dw), contrl (dw),
;                    lcount (dw), pcount, bufseg (dw), bufptr (dw)
;              NOTES: should be used at the start of the CODE listing
;============================================================

setup:
                                            ;==================
                                            ;recover mode and port
                                            ;data
                                            ;==================
        mov     dx,5000h                    ;data cache segment
        mov     bx,0                        ;data cache offset
        push    ds                          ;save data segment
        mov     ds,dx                       ;set data segment to cache
        mov     al,[bx]                     ;get mode value
        inc     bx                          ;offset pointer
        mov     dx,[bx]                     ;get port base address
        pop     ds                          ;recover data segment
        mov     mode,al                     ;save mode value
        inc     dx                          ;offset to status input address
        mov     status,dx                   ;save status port address
        inc     dx                          ;offset to control output
        mov     contrl,dx                   ;save control port address
                                            ;==================
                                            ;set defaults
                                            ;==================
        mov     dx,6000h                    ;default ram buffer start segment
        mov     bx,0                        ;default buffer pointer
        mov     bufseg,dx                   ;save buffer segment value
        mov     bufptr,bx                   ;save buffer pointer
        mov     dx,0                        ;start line count
        mov     cx,128                      ;start pixel count
        mov     lcount,dx                   ;save line count
        mov     pcount,cx                   ;save pixel count
                                            ;==================
                                            ;evaluate mode value
                                            ;==================
scan:
        mov     al,mode                     ;recover mode value
;               *                           ;==================
scanx:
        ret                                 ;exit ml routine
```

```
;===============================================================
;      SAMPLE LISTING: ASM 3
;           CATEGORY: subroutine
;               NAME: chi
;           FUNCTION: checks for HIGH on interface clock
;REGISTERS (ENTRY): doesn't matter
; REGISTERS (EXIT): all preserved
;        SUBROUTINES: none
;             MEMORY: status (dw)
;              NOTES: return executed on clock HIGH
;===============================================================

chi     PROC    NEAR

        push    ax                      ;save registers
        push    dx
        mov     dx,status               ;status port address
chi1:
        in      al,dx                   ;get port data
        and     al,8                    ;mask non-clock bits
        cmp     al,8                    ;clock HIGH?
        jne     chi1                    ;if not, look again
        pop     dx                      ;recover registers
        pop     ax
        ret

chi     ENDP
```

```
;===============================================================
;     SAMPLE LISTING: ASM 4
;          CATEGORY: subroutine
;              NAME: clo
;          FUNCTION: checks for LOW on interface clock
;REGISTERS (ENTRY): doesn't matter
; REGISTERS (EXIT): all preserved
;       SUBROUTINES: none
;            MEMORY: status (dw)
;             NOTES: return executed on clock LOW
;===============================================================

clo     PROC    NEAR

        push    ax              ;save registers
        push    dx
        mov     dx,status       ;status port address
clo1:
        in      al,dx           ;get port data
        and     al,8            ;mask non-clock bits
        cmp     al,0            ;clock LOW?
        jne     clo1            ;if not, look again
        pop     dx              ;recover registers
        pop     ax
        ret

clo     ENDP
```

```
;===============================================================
;     SAMPLE LISTING: ASM 5
;          CATEGORY: subroutine
;              NAME: time
;          FUNCTION: executes a time delay using the interface clock
;REGISTERS (ENTRY): delay value (in clock cycles) in dx
; REGISTERS (EXIT): all preserved
;       SUBROUTINES: chi, clo
;            MEMORY: none
;             NOTES:
;===============================================================

time      PROC     NEAR

          push     dx                  ;save register
time0:
          call     chi                 ;wait for clock HIGH
          call     clo                 ;wait for clock LOW
          dec      dx                  ;decrement delay count
          jnz      time1               ;if not done, wait another cycle
          pop      dx                  ;recover register
          ret

time      ENDP
```

```
;========================================================
;     SAMPLE LISTING: ASM 6
;          CATEGORY: subroutine
;              NAME: getvid
;          FUNCTION: input an 8 bit video value
;REGISTERS (ENTRY): doesn't matter
; REGISTERS (EXIT): all preserved
;       SUBROUTINES: chi, clo
;            MEMORY: status (dw), control (dw), atod (db)
;             NOTES: 8 bit value in atod on exit, time required
;                    is one clock cycle
;========================================================

getvid  PROC        NEAR

        push    ax                  ;save registers
        push    dx
        call    chi                 ;wait for clock HIGH
                                    ;===========================
                                    ;strobe A/D
                                    ;===========================
        mov     dx,contrl           ;control port address
        mov     al,1                ;set STROBE LOW - see text
        out     dx,al               ;send to interface

        nop                         ;delay
        nop                         ;if not done, loop back
        mov     al,0                ;set STROBE HIGH - see text
        out     al,dx               ;send to interface
        call    clo                 ;wait for clock LOW to allow
                                    ;time for A/D conversion
                                    ;===========================
                                    ;get and save high nibble
                                    ;===========================
        mov     dx,status           ;status port address
        in      al,dx               ;get status byte
        and     al,240              ;mask out low nibble
                                    ;===========================
                                    ;invert bit 7 - see text
                                    ;===========================
        cmp     al,127              ;bit 7 HIGH?
        jge     getv2               ;if so, jump around
        add     al,128              ;bit 7 is LOW so set it HIGH
        jmp     getv3               ;jump around
getv2:
        sub     al,128              ;bit is high so set it LOW
getv3:
        mov     atod,al             ;save high nibble
                                    ;===========================
                                    ;toggle nibble select LOW
                                    ;===========================
        mov     dx,contrl           ;control port address
        mov     al,2                ;set AUTO-FEED LOW - see text
        out     dx,al               ;send control byte
```

```
           nop
           nop
                                    ;==========================
                                    ;get low nibble. format 8 bit
                                    ;video value and save it
                                    ;==========================
           mov       dx,status      ;status port address
           in        al,dx          ;get status byte
           and       al,240         ;mask out low nibble
                                    ;==========================
                                    ;invert bit 7 - see text
                                    ;==========================
           cmp       al,127         ;is bit 7 HIGH?
           jge       getv5          ;if so, jump around
           add       al,128         ;bit 7 is LOW so set it HIGH
           jmp       getv6          ;jump around
getv5:
           sub       al,128         ;bit 7 is HIGH so set it LOW
getv6:
           shr       al,1           ;format as low nibble
           shr       al,1
           shr       al,1
           shr       al,1
           and       al,15          ;mask out high nibble
           add       al,atod        ;add high nibble
           mov       atod,al        ;save 8 bit video value
                                    ;==========================
                                    ;reset nibble select
                                    ;==========================
           mov       dx,contrl      ;load control port address
           mov       al,0           ;set AUTO-FEED LOW - see text
           out       dx,al          ;send to interface
           pop       dx             ;recover registers
           pop       ax
           ret

getvid     ENDP
```

```
;=================================================================
;     SAMPLE LISTING: ASM 7
;          CATEGORY: sub-routine
;              NAME: lookup
;          FUNCTION: retrieves value from lookup table
;REGISTERS (ENTRY): doesn't matter
; REGISTERS (EXIT): all preserved
;        SUBROUTINES: none
;            MEMORY: atod
;             NOTES: assumes a/d value is in atod on entry,
;                    new value will be in atod on exit
;=================================================================

lookup    PROC      NEAR

          push      ax                  ;save registers
          push      bx
          push      dx
          mov       al,atod             ;recover a/d value
          mov       ah,0                ;zero high byte
          mov       bx,100h             ;table pointer
          add       bx,ax               ;offset pointer
          mov       dx,5000h            ;table segment
          push      ds                  ;save data segment
          mov       ds,dx               ;set up table segment
          mov       al,[bx]             ;get table value
          pop       ds                  ;recover data segment
          mov       atod,al             ;save converted value
          pop       dx                  ;recover registers
          pop       bx
          pop       ax
          ret

lookup    ENDP
```

```
;================================================================
;      SAMPLE LISTING: ASM 8
;             CATEGORY: subroutine
;                 NAME: wrpix
;             FUNCTION: writes a 4 bit pixel to the screen
;REGISTERS (ENTRY): doesn't matter
; REGISTERS (EXIT): all preserved
;          SUBROUTINES: none
;               MEMORY: atod, pcount, lcount
;                NOTES: pixel and line counts are updated
;================================================================

wrpix     PROC      NEAR

          push      ax                   ;save registers
          push      bx
          push      cx
          push      dx
          mov       cx,pcount            ;get pixel count
          mov       dx,lcount            ;get line count
          mov       bh,0                 ;set up for write
          mov       ah,0ch
          mov       al,atod              ;get video value
          int       10h                  ;write to screen
          mov       cx,pcount            ;get pixel count
          mov       dx,lcount            ;get line count
          inc       cx
          cmp       cx,640               ;end of line?
          jne       wrpixx               ;if not, jump around
          mov       cx,128               ;line start
          inc       dx                   ;increment line count
          mov       lcount,dx            ;save line count
wrpixx:
          mov       pcount,cx            ;save pixel count
          pop       dx                   ;recover registers
          pop       cx
          pop       bx
          pop       ax
          ret

wrpix     ENDP
```

```
;================================================================
;       SAMPLE LISTING: ASM 9
;             CATEGORY: sub-routine
;                 NAME: rdram
;             FUNCTION: reads a value from the ram buffer
;REGISTERS (ENTRY): doesn't matter
; REGISTERS (EXIT): all preserved
;          SUBROUTINES: none
;               MEMORY: atod, bufseg, bufptr
;                NOTES: ram value in atod on exit; segment and
;                       buffer pointers updated automatically
;================================================================

rdram   PROC    NEAR

        push    ax              ;save registers
        push    bx
        push    dx
        mov     bx,bufptr       ;get ram pointer
        mov     dx,bufseg       ;get ram segment
        push    ds              ;save data segment
        mov     ds,dx           ;set up video ram segment
        mov     al,[bx]         ;read the byte
        pop     ds              ;recover data segment
        cmp     bx,0ffffh       ;end of segment?
        jne     rdram1          ;if not, jump around
        mov     bx,0            ;zero ram pointer
        add     dx,1000h        ;update segment
        mov     bufseg,dx       ;save segment
        jmp     rdramx          ;jump around
rdram1:
        inc     bx              ;increment ram pointer
rdramx:
        mov     atod,al         ;save video value
        mov     bufptr,bx       ;save ram pointer
        pop     dx              ;recover registers
        pop     bx
        pop     ax
        ret

rdram   ENDP
```

```
;================================================================
;      SAMPLE LISTING: ASM 10
;            CATEGORY: sub-routine
;                NAME: wrram
;            FUNCTION: writes a video value to ram
;REGISTERS (ENTRY): doesn't matter
; REGISTERS (EXIT): all preserved
;         SUBROUTINES: none
;              MEMORY: atod, bufseg, bufptr
;               NOTES: segment and ram pointers updated automatically
;================================================================

wrram     PROC      NEAR

          push      ax                  ;save registers
          push      bx
          push      dx
          mov       bx,bufptr           ;get ram buffer pointer
          mov       dx,bufseg           ;get ram buffer segment
          mov       al,atod             ;get video value
          push      ds                  ;save data segment
          mov       ds,dx               ;set up ram buffer segment
          mov       [bx],al             ;send video to ram
          pop       ds                  ;recover data segment
          cmp       bx,0ffffh           ;end of segment?
          jne       wrram1              ;if not, jump around
          mov       bx,0                ;reset pointer
          add       dx,1000h            ;update segment value
          mov       bufseg,dx           ;save segment value
          jmp       wrramx              ;jump around
wrram1:
          inc       bx                  ;increment pointer
wrramx:
          mov       bufptr,bx           ;save pointer
          pop       dx
          pop       bx
          pop       ax
          ret

wrram     ENDP
```

```
;================================================================
;    SAMPLE LISTING: ASM 11
;          CATEGORY: main program listing
;              NAME: dump
;          FUNCTION: passes an image from ram to the screen
;REGISTERS (ENTRY): doesn't matter
; REGISTERS (EXIT): doesn't matter
;       SUBROUTINES: rdram, lookup, wrram
;            MEMORY:
;             NOTES: results depend on current look-up table
;================================================================

dump:
        call    rdram           ;read video ram
        call    lookup          ;convert for display
        call    wrpix           ;write to screen
        mov     dx,lcount
        cmp     dx,480          ;last line?
        jne     dump            ;if not, do another
        jmp     scanx           ;exit assembly module
```

```
;================================================================
;     SAMPLE LISTING: ASM 12
;           CATEGORY: sub-routine
;               NAME: pixel
;           FUNCTION: samples a video value and posts it to
;                     the screen and to ram
;REGISTERS (ENTRY): doesn't matter
; REGISTERS (EXIT): all preserved
;        SUBROUTINES: getvid, wrram, lookup, wrpix
;             MEMORY: none
;              NOTES:
;================================================================

pixel   PROC    NEAR

pixel:
        call    getvid          ;sample video
        call    wrram           ;store in ram
        call    lookup          ;convert a/d value
        call    wrpix           ;display on screen
        ret

pixel   ENDP
```

```
;================================================================
;     SAMPLE LISTING: ASM 13
;           CATEGORY: subroutine
;               NAME: wline
;           FUNCTION: samples, stores, and displays a WEFAX
;                     line
;REGISTERS (ENTRY): doesn't matter
; REGISTERS (EXIT): doesn't matter
;        SUBROUTINES: none
;             MEMORY: none
;              NOTES:
;================================================================

wline   PROC    NEAR

        push    cx
        push    dx
        mov     cx,512          ;line sample count
wlr:
        call    pixel           ;sample pixel
        call    chi             ;delay 1 clock cycle
        call    clo
        dec     cx              ;decrement sample count
        jnz     wlr             ;if not last sample, do again
        mov     dx,1024         ;250 mS delay value (1 line)
        call    time
        pop     dx
        pop     cx
        ret

wline   ENDP
```

Chapter 7

High Resolution Direct-Readout Systems

By Dr Jeff Wallach, N5ITU
   Dallas Remote Imaging Group

There has been a significant expansion in the number of direct-readout weather-satellite stations (both amateur and professional) over the past few years. Whereas the number of operational direct-readout stations ranged in the hundreds ten years ago, there are now thousands of APT/WEFAX installations in use on a daily basis by amateurs, educators, scientists, professional meteorologists, military units, and the commercial sectors. The exponential growth of direct-readout imaging systems is directly related to several technology and marketplace issues.

Perhaps the biggest impact on the use of APT and WEFAX ground stations has been the declining costs and complexity of the hardware. Ten to fifteen years ago, most amateur imaging stations were based on the older facsimile machine technology, or used long-persistence phosphor oscilloscopes to build up a poor resolution image ingested from the TIROS and Advanced-TIROS satellites. This equipment, although fairly inexpensive, was difficult to maintain and provided resolution of only 1 to 3 bits (2 to 8 levels of grayscale). These early systems were fun to play with, but provided little in the way of usable data, particularly when temperature calibration, geopolitical boundaries, or complex image-processing was required.

The rapid advances and declining costs in personal computer processing, super VGA display technology, and large (gigabyte) storage capacity have been a prime factor in the advancements made in personal direct readout imaging systems over the past decade. Simple plug-in computer boards, graphical user interfaces, and powerful image processing routines all provided the impetus for a surge in the interest in weather satellite imagery systems for novices and professionals alike. The advancements in graphical satellite tracking/prediction programs, availability of current Keplerian elements on electronic bulletin boards, and excellent support by hardware vendors and outside consultants also added to the widespread distribution of Earth resources imaging systems.

It was inevitable that users of these direct-readout systems would escalate their requirements to the higher resolution data products available from the polar and geostationary orbiting weather satellites. The 4-kilometer resolution polar-orbiting APT images and 4- to 7-kilometer resolution of GOES WEFAX images are in wide use today. However, there is an ever-increasing requirement for the 1.1-kilometer resolution of the NOAA HRPT (High Resolution Picture Transmission) images, as well as the 0.8-km resolution of the GOES VAS (Visual Atmospheric Sounder) and European Space Agency METEOSAT HR (High Resolution) imagery. Over the years, amateurs had dabbled in building these high-resolution ground stations for their personal enjoyment. The cost of commercial systems was simply out of reach for most of the direct-readout community (systems cost in the range of $250,000 up to over one million dollars!).

Through the dedicated work of some of the pioneers in amateur direct imaging (Dr John DuBois, Ed Murashie, Tracy Lenocker, Tom Loebl, Roger Beale, Marciano Righini, Guido Emiliani and others), HRPT, VAS, and HR imaging stations are now available to the direct-readout community. Prices range from $2000 to $10,000—several orders of magnitude less expensive than those available only ten years ago.

This chapter is dedicated to discussing amateur use of the high resolution HRPT data. We'll also take a building-block approach to constructing (from a kit or commercially integrated system) high resolution direct imaging stations. The highly technical background information will not be discussed here, but may be further investigated in the publications referenced at the end of this chapter.

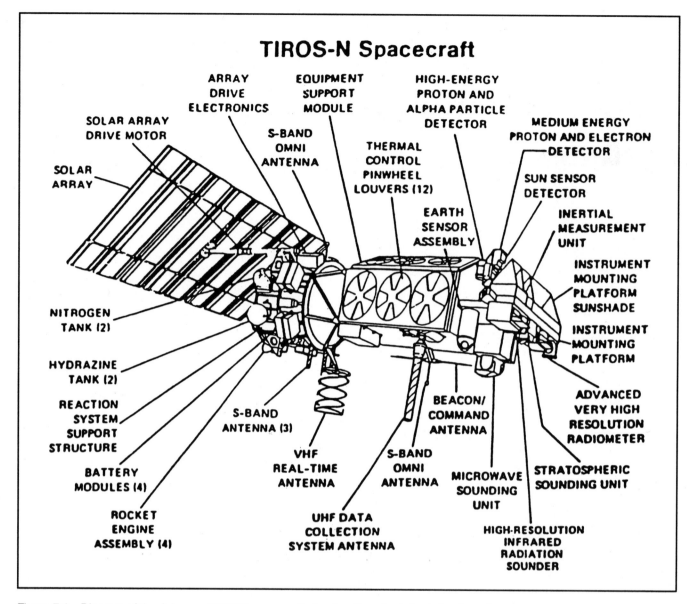

TIROS-N Spacecraft

ARRAY DRIVE ELECTRONICS

EQUIPMENT SUPPORT MODULE

HIGH-ENERGY PROTON AND ALPHA PARTICLE DETECTOR

MEDIUM ENERGY PROTON AND ELECTRON DETECTOR

SOLAR ARRAY DRIVE MOTOR

S-BAND OMNI ANTENNA

THERMAL CONTROL PINWHEEL LOUVERS (12)

SUN SENSOR DETECTOR

SOLAR ARRAY

EARTH SENSOR ASSEMBLY

INERTIAL MEASUREMENT UNIT

INSTRUMENT MOUNTING PLATFORM SUNSHADE

NITROGEN TANK (2)

INSTRUMENT MOUNTING PLATFORM

HYDRAZINE TANK (2)

REACTION SYSTEM SUPPORT STRUCTURE

S-BAND ANTENNA (3)

BEACON/ COMMAND ANTENNA

ADVANCED VERY HIGH RESOLUTION RADIOMETER

BATTERY MODULES (4)

VHF REAL-TIME ANTENNA

S-BAND OMNI ANTENNA

MICROWAVE SOUNDING UNIT

STRATOSPHERIC SOUNDING UNIT

ROCKET ENGINE ASSEMBLY (4)

UHF DATA COLLECTION SYSTEM ANTENNA

HIGH-RESOLUTION INFRARED RADIATION SOUNDER

Figure 7.1—Diagram of the Advanced TIROS spacecraft showing the orientation of major subsystems.

In order to provide a solid framework of understanding, we will review both the polar and geostationary satellite-imaging systems. The lower-resolution APT/WEFAX imagery will be compared to the higher-resolution HRPT/VAS/HR images to illustrate the benefits of more advanced imaging ground stations. Let's start with a review of the Advanced-TIROS (*Tele-vision InfraRed Observation Satellite*).

HIGH-RESOLUTION PICTURE TRANSMISSION (HRPT) IMAGERY ON THE AVHRR INSTRUMENT

The National Oceanic and Atmospheric Administration (NOAA) is responsible for operating the Advanced TIROS series of weather satellites. Most weather satellite enthusiasts are already familiar with one of the products

produced by these polar-orbiting satellites: the Automatic Picture Transmission (APT) imagery. What is not widely understood is that the APT imagery is actually a derivative (byproduct) of the High Resolution Picture Transmission (HRPT) product. The satellite carries an Advanced Very High Resolution Radiometer (AVHRR) instrument that generates *both* the HRPT and APT imagery.

The AVHRR is the latest in a long series of instruments flown aboard NOAA's polar weather satellites. The original service provided by the early weather satellites was actually a television-based system using automatic (analog) picture transmissions. In 1972, a new breed of instrument was developed based on a *radiometer* design. It was called a Very High Resolution

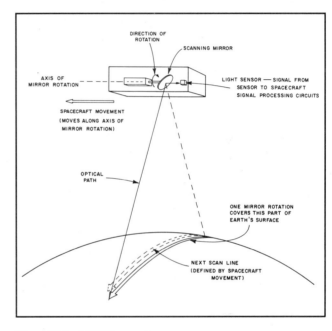

Figure 7.2—The Advanced Very-High Resolution Radiometer (AVHRR) instrument flown on the NOAA 11 spacecraft. (*Photograph courtesy of ITT Aerospace Division*)

Radiometer (VHRR). The instrument measured reflected light and temperature, passing the information to ground stations by way of analogue telemetry. Further advances in the late 1970s produced the *digital* Advanced Very High Resolution Radiometer. These early AVHRR instruments employed four "channels," or spectral bands, with up to 10 bits (1024 gray levels) of radiometric resolution in the weather satellite image. In June 1981, a new AVHRR/2 instrument was flown that had five channels available for both visible (one channel), infrared, and near infrared channels (the other four channels).

Figure 7.1 shows the system diagram of the Advanced TIROS satellite and indicates the position of the AVHRR instrument. Figure 7.2 is a photograph of the AVHRR/2 package that flew aboard NOAA 11.

The AVHRR instrument utilizes a $45°$ scanning mirror that rotates at 360 rpm from an altitude of 870 km. The scanning track across the Earth measures both the visible reflected light and infrared wavelengths, building an image in the process. The satellite motion across its orbital track causes the successive scan lines to form a contiguous, two-dimensional image. The total image is composed of 2048 pixels by however many lines can be received by a fixed Earth station that is tracking the satellite as it moves across the horizon. Each pixel size in the image is defined to be 1.08 km at the nadir point. As the image moves away from the downward nadir point, the pixels get progressively elongated or distorted. There are six lines transmitted per second. A schematic diagram of the scanning process is shown in Figure 7.3.

The scanning mirror itself is elliptically shaped and approximately 8.25 inches by 11.5 inches. The mirror is large enough to fill the field of view of the telescope lens, which is approximately eight inches in diameter. The optical system splits the incoming light into discrete spectral bands focused on two electronic detectors. These detectors are sensitive to visible, infrared,

Figure 7.3—HRPT scanning

and near-infrared wavelengths. Visible and near-infrared channels utilize detectors made out of silicon, with the infrared sensors made of various compound metals. When the light falls on these detectors, it generates a proportional electric current which is amplified and converted to digital information.

As mentioned, the AVHRR instrument provides a 1.1-kilometer resolution directly under the satellite. Due to the geometric distortion that occurs away from nadir, the resolution drops to about 4 kilometers at the leading and trailing ends of each scan line (this distortion is noticeably visible in the image). The Automatic Picture Transmission (APT) data, which is in wide use today, is actually derived from the output of the AVHRR instrument. Basically, two of the five channels

Figure 7.4—A visible HRPT image of southern Europe (1.1-kilometer resolution) as received in Italy by Marciano Rhigini.

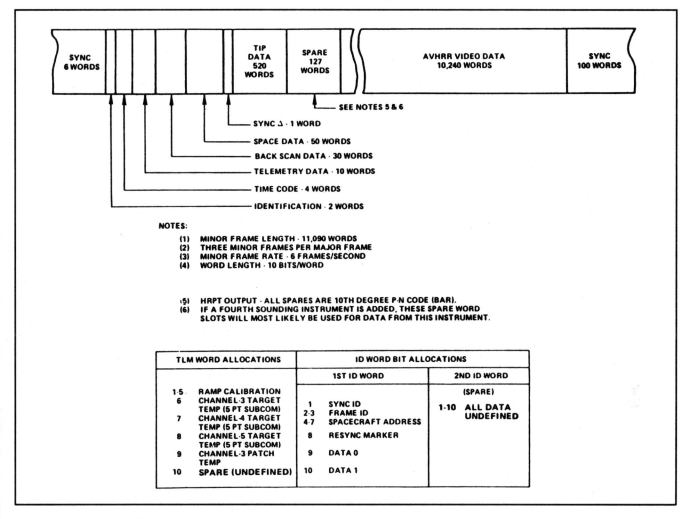

NOTES:

(1) MINOR FRAME LENGTH · 11,090 WORDS
(2) THREE MINOR FRAMES PER MAJOR FRAME
(3) MINOR FRAME RATE · 6 FRAMES/SECOND
(4) WORD LENGTH · 10 BITS/WORD

(5) HRPT OUTPUT · ALL SPARES ARE 10TH DEGREE P-N CODE (BAR).
(6) IF A FOURTH SOUNDING INSTRUMENT IS ADDED, THESE SPARE WORD
    SLOTS WILL MOST LIKELY BE USED FOR DATA FROM THIS INSTRUMENT.

TLM WORD ALLOCATIONS		ID WORD BIT ALLOCATIONS	
		1ST ID WORD	2ND ID WORD
1-5	RAMP CALIBRATION		(SPARE)
6	CHANNEL-3 TARGET TEMP (5 PT SUBCOM)	1 SYNC ID	1-10 ALL DATA UNDEFINED
7	CHANNEL-4 TARGET TEMP (5 PT SUBCOM)	2-3 FRAME ID	
8	CHANNEL-5 TARGET TEMP (5 PT SUBCOM)	4-7 SPACECRAFT ADDRESS	
9	CHANNEL-3 PATCH TEMP	8 RESYNC MARKER	
10	SPARE (UNDEFINED)	9 DATA 0	
		10 DATA 1	

Figure 7.5—The HRPT data frame format.

are chosen and *every third line* is transmitted, thus the two-line-per-second scan rate (or more commonly, 120 lines per minute). The difference is that the APT image is corrected for geometric distortion by averaging a variable number of adjacent pixels along each line. This averaging yields about a 4-kilometer image resolution. The minute markers and telemetry data are added and the signal is broadcast on 137.50 or 137.62 MHz.

The actual AVHRR instrument contains several electronics modules, including:

- An S-band transmitter (1698 MHz) for the real-time transmission (direct readout) of the AVHRR image data and satellite housekeeping telemetry
- Two more S-band transmitters (1707 and 1702.5 MHz) for delayed data transmissions from onboard digital tape recorders.

- A 137-MHz transmitter for lower-resolution APT images (intended for reception by modestly equipped ground stations).

The power output is on the order of 5 W. The HRPT data is transmitted at a rate of 665.4 kilobits per second (kbps) as a split-phase encoded, phase-modulated signal. The signal is transmitted at 1707 MHz (for NOAA 9 and 11), and 1698 MHz (NOAA 10 and 12). A third channel on 1702.5 MHz is available for backup purposes. A summary of the downlink transmitter characteristics is shown in Table 1.

Now that some of the AVHRR characteristics have been discussed, just how good are the HRPT images relative to what we are used to in the APT environment? Figure 7.4 shows a visible-light HRPT image of southern Europe as received in Italy by Marciano Rhigini. The resolution along the satellite ground track is 1.1 kilometers. This spectacular image is rep-

Table 1
Advanced TIROS AVHRR Downlink Characteristics

Orbit:	Polar, sun-synchronous
	450 miles (870 km nominal)
Frequency:	1698, 1707, 1702.5 MHz
Frequency stability:	$\pm 2 \times 10^{-5}$
Transmitter power:	5 W
EIRP:	39.0 dBm
Antenna polarization:	1698, 1707 RHCP
	1702.5 LHCP
Spectrum bandwidth:	<3 MHz
Modulation type:	PCM/PSK $\pm 67°$
Modulation code:	HRPT—Split phase

resentative of the superior resolution attainable with HRPT!

It should be noted that the AVHRR instrument is not the only package on the Advanced-TIROS bus. Other instruments and data products include the TIROS Operational Vertical Sounder (TOVS), the Space Environment Monitor (SEM), the Data Collection System (DCS), and the spacecraft telemetry (TIP) instruments. (See Figure 7.1 for the location of these instruments on the satellite platform.) Most amateurs are typically interested in the weather images, not these other instruments and their associated data products. If you are interested, you can learn more by obtaining a copy of NOAA Technical Memorandum NESS 95, "The TIROS-N/NOAA A-G Satellite Series."

HIGH RESOLUTION PICTURE TRANSMISSION (HRPT) DIGITAL FORMATS

Prior to discussing the basic components of an HRPT ground station, let's review the characteristics of the HRPT downlink data. We'll compare and contrast the HRPT imagery processing requirements with the more common APT imagery.

The HRPT data format is digital, as compared to the analog tones used for APT imagery. The digital format places more stringent requirements on the electronics required to receive and demodulate the downlink, but it is fairly straightforward. Commercial kits and fully assembled/integrated units are available from several vendors.

The HRPT data is formatted at six lines per second. Each line of the HRPT image data contains 11,090 *words* of information (each of these words is 10 bits long—or 1024 levels of grayscale). The Most Significant Bit (MSB) is transmitted first. The actual image portion begins with word location (offset) 751 and is completed at word 10,990. The frame format for the HRPT data is shown in Figure 7.5.

HRPT GROUND STATION REQUIREMENTS

The RF hardware, electronics, software, and personal computer requirements for HRPT image reception are not significantly more difficult than setting up an APT or WEFAX ground station. The basic elements include:

- Four-foot parabolic dish
- Feedhorn and quadrature combiner
- Low-noise amplifier (LNA)
- Downconverter
- Phase-locked loop demodulator
- Bit-synchronization board
- Personal computer
- Image decoding/processing software
- Satellite tracking hardware and software

This sounds like a complex system, but it is relatively simple. There are kits and plug-and-play systems available in the marketplace that will suit a variety of technical requirements and budgets.

As we discussed previously, this type of HRPT system would have cost $250,000 and more just a few years ago. Thanks to the pioneering research and design efforts of Dr John DuBois and Ed Murashie, low-cost systems are now available. And this fine engineering work was complemented by the outstanding graphics support packages and image processing code written by Tracy Lenocker, Tom Loebl and other members of the Dallas Remote Imaging Group (see reference at end of chapter for a free introductory kit). It is now possible to construct a fine HRPT ground station for several thousand dollars that will provide outstanding imagery and image processing capabilities. A block diagram of an HRPT ground station is shown in Figure 7.6 (based on the DuBois design).

Antenna and Feedhorn

In amateur systems, the minimum antenna size for HRPT S-band imagery is a 4-foot (1.2-meter) parabolic dish. Such a dish would realize a gain of around 24 dB. Some amateurs have attempted to use a loop Yagi design for the antenna system (gain less than 22 dB), but the results have not been very satisfactory. A higher noise figure preamp could be used with a larger antenna, but this is not very desirable. The smaller beamwidth and high wind loading are serious disadvantages. The 4-foot dish can be easily steered and has a much lower wind loading factor.

Dishes are commercially available. From a durability, accuracy, and maintainability perspective, they're a wise investment. Homebrew dishes are feasible, or old UHF TV antenna dishes may also be employed.

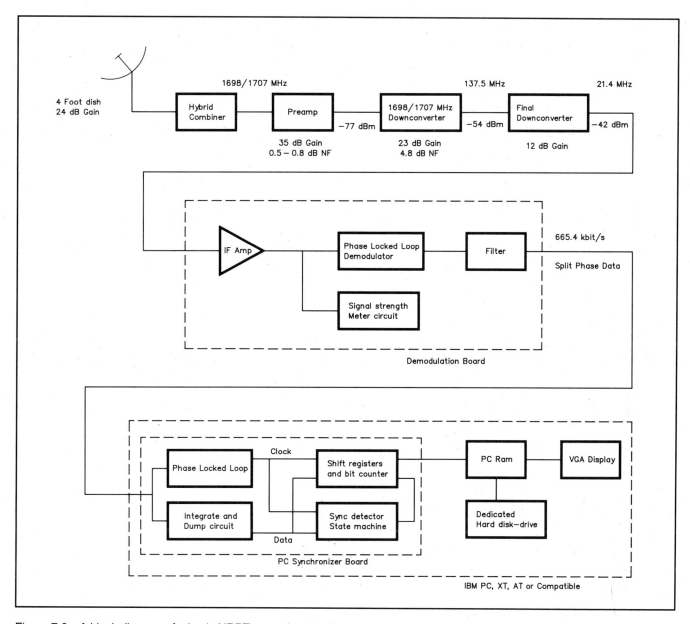

Figure 7.6—A block diagram of a basic HRPT ground station based on design work by John Dubois.

The feedhorn design is another critical component. It is best to purchase a commercially available unit, but feedhorns can be constructed from scratch. The DuBois design utilizes one almost identical to the WEFAX coffee can feed described in Chapter 2. It can be fabricated out of a 26-ounce coffee can, brass sheets or an aluminum can. The major difference is that the HRPT must use a circularly polarized antenna. Two feedhorn probes are needed for the WEFAX design. They must be placed 90° apart from each other around the outer diameter. The coax from the two antenna probes must then be combined in a *quadrature combiner*, also known as a 90° hybrid. This particular design may be purchased from Mini Circuits Labs

(ZAPD-2). The coax feed should be kept very short with the combiner placed just behind the feedhorn. Its single-output coax connects directly to the LNA, which is also just behind the feedhorn. The polarization should be right-hand circular.

Low-Noise Preamplifier (LNA)

One of the most critical aspects of an HRPT downlink station is the sensitivity and noise figure of the preamplifier. This has traditionally been one of the most difficult components for the amateur to build or buy. Data reception with practical-size antennas is not possible unless the LNA performance is extremely good. The LNA noise figure should be 0.8 dB or lower (which corresponds to a noise temperature of 59 Kelvins, or lower) and the gain should be at least 30 dB. The preamp sets the noise figure for the entire system. An LNA with a noise figure under 1 dB, but 20 with dB gain or less, is *not* adequate. It will not overcome the noise of the downconverter unless the downconverter has a noise figure of about 1.5 dB or lower.

With an analog system such as APT or GOES WEFAX, a poor signal-to-noise ratio will result in some image reception (albeit a noisy image). However, the digital HRPT downlink will *not* be properly detected with a poor signal-to-noise ratio due to the threshold effect. The good news is that there are excellent commercial LNAs now available for a reasonable cost. They deliver excellent performance, noise figures, and overall gain.

Downconverter

The preamp is connected to the downconverter with very low loss coax, or it is fully integrated into the downconverter. Fully integrated units are known as *block downconverters*. They feature RF band passes of 16 MHz or less, and convert the 1698- or 1707-MHz signal to a more usable frequency of 133 or 142 MHz (or 128 to 145 MHz for some commercial systems).

Regardless of which type of downconverter is in use, channel selection is accomplished by tuning the intermediate frequency (IF) using phase-locked loop circuitry (PLL). The IF will have a narrower band pass (3 MHz) intended to set the final system noise bandwidth. RF input filtering is very critical at this stage since it determines the susceptibility to out-of-band interference. Typically, the low-end amateur systems use a 2-pole (12-dB/octave) filter, with ultimate rejection as poor as 25 dB to as good as 60 dB, depending on the quality of the unit. Some of the better units have a 4-pole (24-dB/octave) filter and over 60 dB ultimate rejection capability. Top-of-the-line commercial downconverters use 6-pole filters.

The final downconversion stage brings the VHF signal down to a convenient frequency for PLL demodulation. A typical choice for amateur units is 21.4 MHz, which is a common IF frequency for microwave equipment. It can range from 10 to 70 MHz as long as the demodulator PLL is tuned to the same frequency.

Demodulator Section

The next major section in an HRPT ground station is an amplifier-filter to raise the signal level and establish the final system band pass for the phase demodulation. About 10 to 15 dB of gain is required with a bandpass of about 3 MHz. These figures provide a −30 dBm signal to the PLL demodulator. Typical amateur designs use an amplifier, discrete phase-locked-loop phase demodulator, an output filter and a signal-strength meter. The output of this section is a split-

phase modulated signal. Several of the commercial units have all these components fully integrated into one package.

Bit and Frame Synchronization

The final system components are the bit and frame synchronizer boards. On the basis of the DuBois design, this circuitry extracts the 0.665-Mbps clock rate and provides some means of *decommutating* the telemetry and different wavelengths of the image (four or five channels, depending on the satellite AVHRR instrument). Several designs exist for the hardware decommutator circuit. Its purpose is to separate data from one of the HRPT channels before sending it to the computer for display.

The bit-synchronizer and DMA port board take the HRPT data (split-phase signal) from the demodulator board. The clock is derived by a phase-locked loop that ignores the data transitions. The data is extracted by means of a digital integrate-and-dump technique. Frame synchronization is accomplished by a *state machine*, which looks for a 60-bit pattern at the start of every HRPT minor frame. The minor frame contains one scan line of 2048 pixels of the four- or five-channel AVHRR data. The clock and data signal also go to a series of shift registers which form the byte or words that are brought into the personal computer by direct memory access techniques. Several variants of these circuits are used in the commercial systems, but the overall design requirements are similar.

Personal Computer

Perhaps the simplest block of the HRPT ground station is the personal computer. In the mid-1980s, when the overall system was designed, IBM PC/XTs, ATs, and a few 80386 machines were available. Processing power is important to allow data ingestion, formatting, and storage to the hard drive. Hard drive capacities and processing power have increased by an order of magnitude. A typical HRPT 15-minute satellite pass, storing all five channels of data, can result in a file over 100 megabytes in size! This was prohibitive for most stations in the mid-1980s, but it is not uncommon for today's computers to have several-hundred megabyte hard drives. Some have well over a *gigabyte* of magnetic storage available.

A good IBM 486 computer (or compatible) with 8 megabytes of RAM, a 200-megabyte hard drive, a Super VGA graphics and a 17-inch color monitor can be purchased for under $2000. With this type of processing power and data storage available to the direct-readout community, HRPT image reception is now within the realm of the average user. That same personal computer can be used for your APT/WEFAX station (a good base to upgrade to HRPT), satellite tracking, digital image processing programs, and much more.

Video Graphics Display System

Once the image is formatted, the quality of the display is highly dependent on the resolving power of the video-graphics adapter card. This was a major limiting point on the early personal computers, and a big cost factor. Fortunately, most of the newer Super VGA adapter cards for the PCs can display 1024 pixels by 768 lines at 256 levels of grayscale. Some cards will allow $1280 \times 1024 \times 256$ levels of grayscale. The quality of the Super VGA card and monitor will have a significant impact on the displayed imagery, so purchase the best card that your budget will allow. (Very high quality Super VGA cards can be purchased for well under $300 at most discount supply stores.)

COMMERCIAL HRPT SYSTEMS

An HRPT imaging system may seem to be a complex ground station to install and implement. Fortunately, there are several commercial kits and fully integrated systems available for the amateur and professional. These systems can be installed and operational within a few hours!

Quorum Communications of Irving, Texas, has a commercial HRPT system that offers a system that is extremely easy to install and operate. While this system is somewhat expensive for the amateur market, it is a fully integrated system. The Quorum HRPT Data Capture Engine provides all the hardware and software necessary to predict satellite visibility, track the satellite, acquire the data and store it on disk. Once the data is stored on the hard drive, it is available for display with a variety of image-processing applications.

Data from the Quorum Data Capture Engine is saved as multiple frames of 11090 10-bit words right-justified in the PC's native 16-bit word. Additional *pseudoframes* are appended to the file that contains specific satellite and ephemeris information for the captured data. The system consists of the following:

- four-foot aluminum dish (23.7 dBi gain @ 1700 MHz)
- feed-support poles
- dish mount
- heavy-duty AZ/EL positioner
- positioner power supply, cables, hardware
- high-gain LNA (38 dB)
- 1700 MHz feed/downconverter (1693-1710 MHz)
- PC-HRPT AVHRR data receiver
- PC-HRPT AVHRR bit/frame synchronizer
- SatTracker IPC tracking controller
- integrated real-time satellite tracking and data-capture software

The advantages of a commercial system such as the Quorum HRPT Data Capture Engine include:

- Full integration of satellite tracking/image ingestion software
- Better input RF filtering with 6-pole filters (24-dB/octave)
- Better IF output filtering
- SAW filters in IF section to reduce out-of-band interference
- Better shielding between component sections

Other vendors such as TimeStep also produce HRPT kits (based on the DuBois/Murashie design) that allow you to provide the systems integration and build much of the antenna hardware and required positioning from scratch. This is an economical alternative, but the quality will not be as good as commercial systems.

BASIC HRPT IMAGE RECEPTION AND PROCESSING

The AVHRR instrument onboard the spacecraft sends the imagery down in real-time (direct readout) to any ground stations within receiving range of the satellite. The signal is transmitted at 1698 or 1707 MHz at 665.4 kbps in the digital format described above. The S-band transmission from the NOAA satellites requires a suitable antenna, typically a 4- to 6-foot parabolic dish. Dishes larger than this diameter are difficult to track mechanically, and have a very narrow beamwidth. (Tracking is absolutely necessary, and many amateurs and commercial systems use a PC satellite tracking/prediction program to drive an azimuth/elevation antenna rotator.)

Once the S-band signal is received, it must be amplified by the LNA and then downconverted to a more usable frequency. (Signal loss due to long coax runs is considerable in the S-band frequency range.) The signal is downconverted to an IF in the 10- to 70-MHz range. Downconversion allows conventional phase-demodulation techniques and components to be used.

The HRPT data is phase modulated onto the RF carrier with a swing of $\pm 67°$. By limiting the modulation to less than $90°$, the demodulation process is simplified. A well-designed phase-locked loop (PLL) circuit can lock onto the HRPT downlink signal and deliver the split-phase bit stream from the phase detector (see Figure 7.5).

The bit stream must be separated into the clock and data components by synchronizing a local clock to the clock rate of the satellite transmitted data. The local clock is used to process the image data and strip away the split-phase (Manchester) encoding.

Figure 7.8—Visible HRPT image of the West Coast of California, including the San Francisco Bay area. Snow-cover in the Sierras is plainly evident.

Following the bit synchronization and split-phase decoding, the data is processed further to identify the start of each scan line, and to separate the different wavelength channels which have been commutated together. (Depending on the spacecraft AVHRR instrument, there are either four or five spectral channels available.) The essential part of this additional processing is to recognize the fixed bit pattern that identifies the start of each scan line. This 60-bit pattern is referred to as the *frame sync* and is the first 60 bits of the HRPT minor frame format. This synchronization pattern also assists in the identification of the first 10-bit HRPT word. The image processing software can count off each of the 10-bit words following the sync mark, thus allowing the desired HRPT image pixels to be saved in the proper format required by the user's graphical display system. In the case of a personal-computer display, the desired image pixels can be copied into banks of RAM and/or video-adapter memory.

The John DuBois design provides an interface to conduct all of the processing following the demodulator. It offers bit synchronization, decoding, and synchronization. The system collects all of the data into the computer via direct memory access (DMA). The software stores the desired spectral channels on the hard drive during an actual satellite pass. The image can then be displayed on a Super VGA graphics screen using a variety of image-display programs.

DIGITAL HRPT IMAGE PROCESSING

Once the HRPT imagery is ingested and stored on the hard disk, you will want to display the image and perform some processing on the pixels. Commercial

Figure 7.7—An infrared image from NOAA 11 of Italy and the Mediterranean Sea. Note the snow on the Alps to the north.

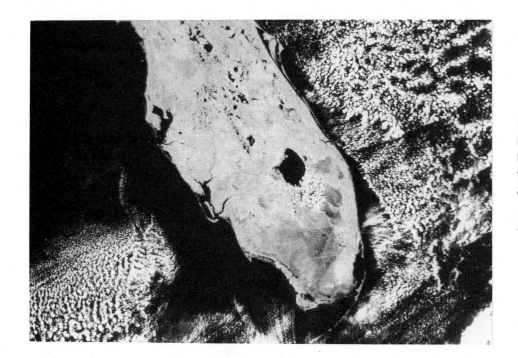

Figure 7.9—A visible HRPT image of southern Florida taken by the AVHRR instrument on the Chinese Feng Yun 1-2 polar orbiter (*captured by the Dallas Remote Imaging Group only 20 orbits after launch*).

HRPT systems integrate image ingestion, satellite tracking, and image display, along with some limited image-processing capability. There are several excellent image-processing programs available (some are free of charge in the public domain) that can manipulate the HRPT pixels.

The advantage of image processing is the ability to enhance specific details in clouds, land features, temperature differences, edge detection, contract enhancement, false colorization, and so on (see Chapter 10). A visible image taken at dawn or dusk may not have very good contrast in the original data. Since the image exists as digital data in a file, it is quite simple for the PC to manipulate these pixels to shift the darker pixels into a brighter pattern. Contrast may be enhanced as well. These newly processed images can

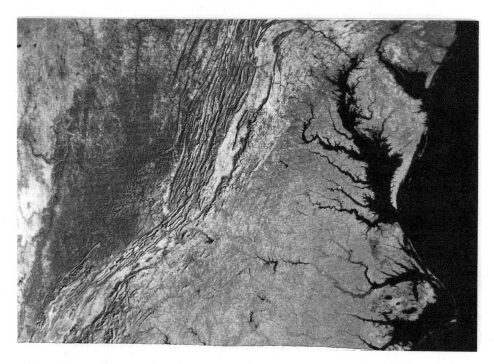

Figure 7.10—A visible-light HRPT image of the Appalachian Mountains and Washington, DC. (also from Feng Yun 1-2).

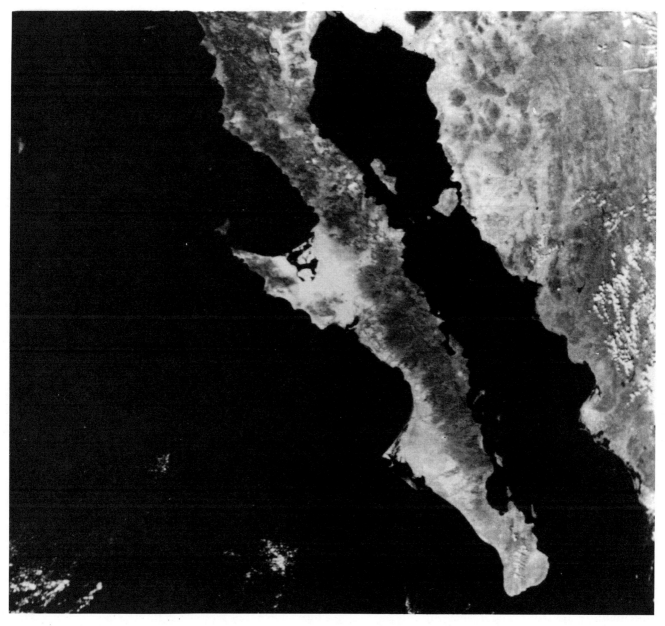

Figure 7.11—Visible-light HRPT image of Baja California and adjacent mainland Mexico.

then be saved to a different file. Several images can be combined in a *loop* to show animation of cloud formations, hurricane tracks, and shifts in ocean currents. The possibilities are endless.

One very excellent and inexpensive digital image-processing program is called SATVIEW. SATVIEW was developed by Tracy Lenocker and Ed Murashie, and was designed specifically to enhance satellite imagery. SATVIEW allows you to view and manipulate all five channels of an HRPT image. It contains many digital image-processing algorithms to enhance the APT and HRPT satellite imagery. SATVIEW will handle a variety of file formats including raw, 8-bit binary, GIF, TIFF,

and so on. Images can be enhanced and colorized, then saved as a GIF image and imported into other graphics applications. Images may be filtered using convolution matrices for softening, gentle sharpening, on-screen histogram display, zooming in and out, brightness and contrast control, enhancement curves, noise removal and rotation. A demonstration copy of SATVIEW may be downloaded from the Dallas Remote Imaging Group Bulletin Board at 214-394-7438.

Other image display programs are also available on the DRIG BBS, including VPIC, CSHOW, ALCHEMY, and PHOTOLAB. All of these programs can view and manipulate weather satellite images. One common image format is the GIF file (Graphics Interchange File) which may be viewed on IBM PCs, Apple Macintoshes, Commodore Amigas, UNIX computers and others. It is easily ported to almost any operating system. GIF has become a very popular format, and many weather satellite imaging systems can generate GIF file formats. There are several commercial graphics applications available that also import GIF file formats. Some of these programs are capable of animating satellite images, thus being able to show weather movement over time as is done on commercial TV weather forecasts. See Figs 7.7 through 7.11 for spectacular examples of HRPT imaging.

VISIBLE ATMOSPHERIC SOUNDER (VAS) IMAGES FROM GOES SATELLITES

Another source of global cloud imagery is the Visible and Infrared Spin-Scan Radiometer Visible Atmospheric Sounder (SVAS) transmissions from the GOES geostationary orbiting satellites. These images are scanned at 22,300 miles above the equator, but still provide some excellent resolution and imagery on a direct-readout, 24-hour-per-day basis.

Previous chapters have already discussed the polar-orbiting satellites that provide the low resolution APT and high resolution HRPT imagery from a sun-synchronous orbit of about 870 kilometers. The GOES spacecraft also provide both low and high resolution weather satellite imagery, but from a much higher, geostationary orbit. NOAA operates both the GOES East and GOES West satellites, while the European Space Agency has lent the US the METEOSAT 3 geostationary spacecraft to augment the global view from space.

A geosynchronous satellite maintains its position at a fixed point above the equator, viewing nearly half of the globe at any one time. This enables the satellite to produce frequent high- and low-definition imagery to complement the polar-orbiting imagery products.

The United States pioneered the use of geostationary satellites with the ATS program, first launched in 1966. This imaging satellite was followed by several successors, including the SMS program and the GOES series which commenced in 1975.

The primary imaging instrument is the visible and infrared spin-scan radiometer (VISSR). The instrument was upgraded on GOES 4 with the incorporation of an atmospheric sounding facility. This instrument was called the VISSR Atmospheric Sounder, or VAS. This radiometer has a similar function to the polar-orbiting AVHRR instrument. It is capable of delivering an absolute estimate of the light reaching it, to a reasonable level of accuracy. It measures both visible and infrared frequencies and provides 24-hour-per-day coverage of an entire hemisphere.

The GOES spacecraft utilize the VISSR instrument by spinning the entire satellite at about 100 rpm. The gyroscopic effect of spinning improves the stability of the satellite considerably, making it easier to design a scanning instrument to take stable imagery of the Earth. Part of the vehicle is despun to allow the transmitting antenna to beam at a fixed point on the ground. Other instruments onboard the GOES spacecraft include an Earth-field magnetometer, solar X-ray sensor, and particle detectors. In addition, a transponder system collects data from remote instrument platforms on the Earth and retransmits it to central processing points. Figure 7.12 shows a schematic diagram of the GOES spacecraft and the location of the VAS instrument. A good review of the GOES satellites and instruments is found in the *GOES D,E, F Data Book* published by NOAA.

The VAS images are created by light from the Earth falling on a vertical scanning mirror. It then passes through a mirror optical system and a set of filters that focus the energy on the respective visible and infrared sensors. The scanned pattern resembles a television image.

West-to-east scan lines are formed by rotation of the spinning spacecraft. The lines are moved by stepping the mirror to scan north to south. Synchronizing signals are also sent through the spacecraft so that imaging signals can be assembled on the ground. The spacecraft spin is 100 rpm or 600 ms per horizontal scan. After each revolution, the vertical mirror is stepped to bring the next strip of Earth into view of the sensors. There are 1820 steps of this mirror so the entire image takes 18.2 minutes to scan.

There are multiple sensors at the focal point of the optical system. At each sampling instant of the horizontal scan, two infrared and eight visible samples are taken. Visible data are detected and converted to electrical signals by an array of eight photomultiplier tubes. Each tube views an eighth of the scan line width, so that eight lines are scanned on each spin of the spacecraft. Resolution at the subsatellite point for each pixel in the image is about 0.8 kilometer. For the infrared, VAS has a total of six detectors. Two are long-wave detectors primarily used for the imaging products. They have a resolution for each IR pixel of approximately 6.9 kilometers.

A digital multiplexer takes the eight visible light samples, the two infrared samples, and various other samples of telemetry data and formats them into a digital data stream at 28 megabits per second (mbps). The data is four-phase modulated (QPSK) on a carrier at 1681.6 MHz and transmitted to Earth through a small 18-dB gain parabolic antenna. As mentioned previously, the antenna is despun or fixed in attitude while the rest of the spacecraft spins under it so that it can be precisely pointed at the Earth.

CURRENT NOAA GOES SATELLITE
(Points toward Earth)

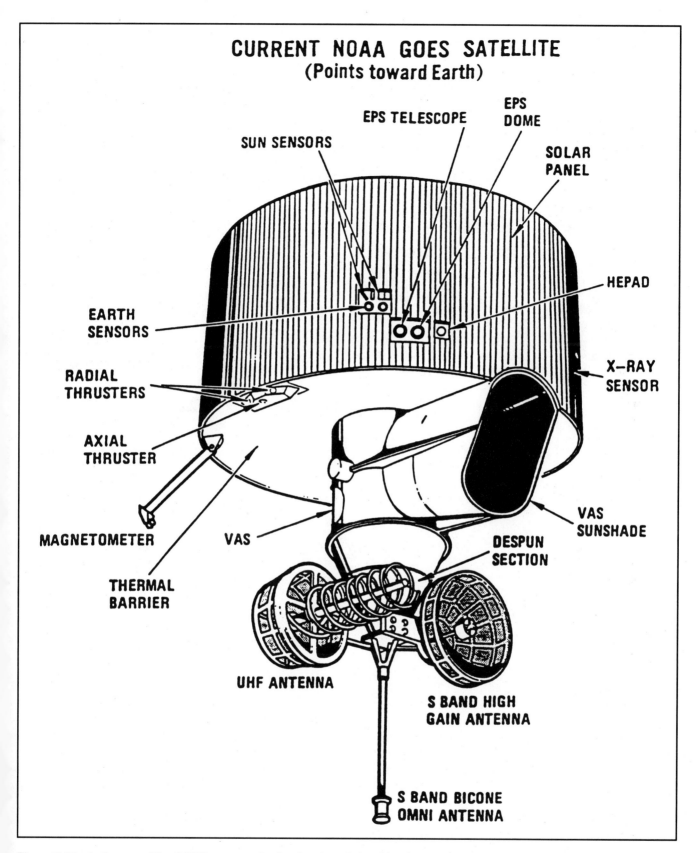

Figure 7.12—A diagram of the GOES spacecraft, showing the relationship of some of the major operational systems.

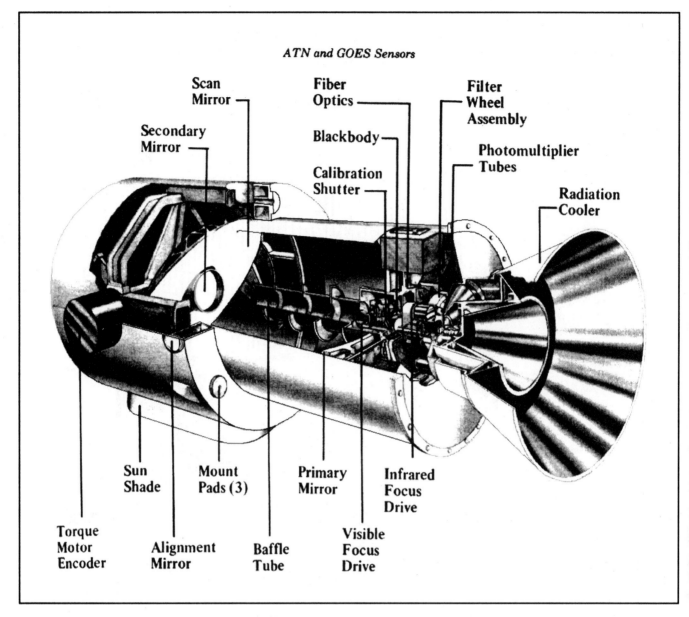

Figure 7.13—Schematic diagram of the GOES visible/infra-red spin-scan radiometer (VISSR).

The transmission of 28 Mbps QPSK data at 1681.6 MHz is quite difficult to receive. So difficult, in fact, that it was never planned in the overall GOES mission for direct readout to ground station users. Instead, a more "user-friendly" data product called s*tretched VISSR* was designed into the system. This is essentially the same data at a lower data rate (2.11 Mbps) which greatly eases the burden on direct-readout users. The raw 28-Mbps data is actually received at the Wallops Island ground station, processed, formatted into the VAS AAA format, and retransmitted back to the satellite for dissemination to ground stations at 2.11 Mbps! Stretched VISSR, or more correctly now, stretched VAS, is the signal that a handful of professional, gov-

ernment, military, university, and commercial users around the world pick up to obtain high resolution digital image data from GOES.

The details of how image samples are formatted for transmission by SVAS will be described later in this section. The data stream transmitted on 1687.1 MHz is modulated onto the carrier as bi-phase modulation (BPSK), 0° and 180°. The data bit rate is 2.11 Mbps. The spectrum of this signal is wide compared to the lower-resolution 1691-MHz WEFAX product. It consists of a broad spread of power between the nulls at the carrier frequency, $\pm N$ times the bit rate. Its power spectrum continues on either side of a peak in a series of nulls. There is a visible notch at the exact middle of the SVAS

Table 2
Basic Specifications for the GOES SVAS System

Orbit:	Geostationary (22,300 mi)
Frequency:	1681.5 MHz (1687.1 SVAS)
Transmitter power:	20 W
EIRP:	55 dBm
Bandwidth:	8 MHz SVAS
Modulation code:	QPSK (1681 MHz)
	BPSK (1687.1 MHz)
	NRZ-L

spectrum. This is the carrier frequency point but the notch confirms a theoretical property of BPSK modulated signals . . . the carrier is absent. This gives BPSK high efficiency; power is not wasted in a transmitted carrier. It is necessary to include at least the central lobe, which is 4 MHz wide at this point, and preferably the first two side-lobes, which require 8 MHz in the receiver band pass, in order to preserve signal quality. Quality is expressed as the signal-to-noise ratio (S/N), and is precious in a SVAS system. It must be preserved at every possible point for good imagery.

A schematic diagram of the VISSR instrument on board the GOES satellites is shown in Figure 7.13. The basic specifications for the GOES SVAS system are shown in Table 2.

The SVAS data format is organized according to a scheme called "MODE AAA." This format specifies the exact placement of telemetry data, image data, and synchronizing bits within the overall flow.

NOAA is in the process of revising this formatting scheme to another process called "GVAR" (GOES VARiable). When it is operational, ground station SVAS equipment will need to be redesigned and reconfigured for the new GVAR format.

Data Format

The basic frame of stretched VAS data is organized as 12 blocks, each lasting 50 milliseconds. The entire frame, which is 600 milliseconds long, is transmitted during one revolution of the satellite spinning at 100 rpm. The organization of the data in the 12 blocks of each line is described in the NOAA/NESDIS publication *Operational VAS MODE AAA Format, Specification SFP 002—Version 2.2*, obtainable from NOAA/NESDIS in Washington, DC.

Block Contents

Each block consists of three segments: a synchronization phase, a header, and the information fields.

The beginning of each block is identified by a unique 10,032-bit pattern in the data. The frame synchronizer circuitry must identify a fragment (or all) of this pattern precisely. It must then use the identification time as a reference point from which to count out bits and words representing the desired weather image.

The general-information header gives the block number and other details. It also includes an information field containing image documentation and the image data itself. The frame of 12 blocks contains data for one line of infrared image in two different spectral bands and data for eight consecutive lines of visible spectrum image. There are 3822 10-bit image words in the infrared block and 15,802 6-bit image words in the visible-spectrum blocks.

The documentation fields of these blocks contain a wealth of information for calibration, gridding, location of the field of view, orbit parameters, and so forth. With so much information available to the ground station user, it is important to decide what data is required and what information may be thrown out. The DuBois design handles this information overload by selecting a small portion (512 bytes) of the incoming data (which is adequate for user purposes) and saving only that portion in a cache RAM on the circuit board. At the end of the block there is adequate time to transfer this cache into the computer.

By saving only every 30th image word, for example, the entire width of the full Earth globe can be displayed in a 512-byte image line. The actual resolution in this image would be quite low. When smaller geographic areas are desired, however, fewer image words are omitted and the resolution is approximately matched to the field of view.

Now that we have an idea of the complexity of the data relative to the simple, analog, AM modulation of the WEFAX imagery, let's take a look at the hardware required to receive SVAS imagery.

GROUND-STATION COMPONENTS OF A VAS RECEIVING SYSTEM

The components of an amateur VAS receiving station include:

- Parabolic dish: 12 to 16 feet
- LNA/Feedhorn: 0.25 dB noise figure, 38-55 dB gain
- Downconverter: 1687/70 MHz (12-15 dB gain)
- 70-MHz IF amplifier: 50-70 dB gain, 8-10 MHz bandwidth
- Demodulator and bit frame synchronizer
- Personal Computer: 8- to 24-bit SVGA display
- Capture and display software

Figure 7.14—A visible-light image of Hurricane Andrew, which devastated the Florida coast in 1992. This image was obtained by GOES from its geostationary vantage point, 22,300 miles above the earth.

Because of the high frequency, digital nature of VAS, building a ground station from simple components is not an easy task for the amateur. Thus, as of early 1993, there are not as many kits or inexpensive commercial systems available. Once again, however, Dr John DuBois and Ed Murashie have designed a system that is feasible to construct for under $2000, excluding computer systems. The primary components of the DuBois design include:

- 12-16 foot parabolic dish
- Feedhorn based on Taggart WEFAX design
- DuBois/Ehrler LNA design
- Quorum downconverter
- Demodulator circuitry
- Phase-locked loop Costas circuit

- Bit synchronizer circuit
- Clock recovery circuit
- Frame Synchronizer circuit
- Personal computer
- SVGA video-display system
- Customized image ingest and display software

This design is not currently in the commercial market, but Dr DuBois has had several write-ups on the basic design requirements and recommendations for implementation.

The SVAS systems have software for image processing, selection and storage. Once the images are displayed and stored digitally, the various image processing programs previously mentioned

Figure 7.15—A rare shot! The moon appears in the field of view of the GOES VISSR instrument as it routinely scans the Earth below.

(SATVIEW, IMDISPLAY, IMPROCES, etc) may be used to manipulate the image pixels as required.

Several vendors are now designing VAS kits and/or integrated systems. Check the DRIG bulletin board system for updates on commercial vendors and kits available for VAS.

Several amateurs have questioned which system they should focus on, HRPT or VAS. This question really can only be answered based on the mission of the ground station and data products required.

The primary advantages of the VAS system would include no requirements to track the satellite (since it is fixed in space as a geostationary platform), visible resolution close to the HRPT resolution, and 24-hour-per-day direct readout. It should be noted, however, that although the visible image resolution is the same as HRPT, the distance and geometric distortion render

images that are not as sharp as AVHRR pictures from 450 miles in space. Figures 7.14 and 7.15 illustrate the capabilities of VAS imagery from the GOES spacecraft.

THE METEOSAT DIGITAL SYSTEM—A PRIMARY DATA USER STATION (PDUS)

The European Space Agency (ESA) operates their own series of geostationary meteorological satellites, called METEOSAT. This series of satellites has a low resolution direct readout similar to WEFAX (Secondary User Data Station—SDUS), and a high resolution image product called High Resolution (HR), designed for the Primary User Data Station (PDUS). There are currently 6 METEOSATS in geostationary orbit.

With the failure of the primary imaging bulbs on one of the two GOES satellites, NOAA asked the European Space Agency to "lend" the US one of the

Table 3
Main Characteristics of the METEOSAT Series Satellites

Orbit:	Geostationary
Data types:	Visible, IR, Water Vapor (Analog and Digital)
Frequencies:	1694.5, 1691 MHz
Modulation:	Analog—AM/FM (2400 Hz) Digital—PCM
Bit Rate:	Digital—166 Kbps MOP—333 Kbps
Line Rate:	Analog—240 lines/min. (SDUS)
Primary Instrument:	Scanning Radiometer (3 Channels — Visible, IR, Water Vapor) All 8-bit data (256 levels)

METEOSAT satellites to assist the GOES east area of coverage. This was critical for the tracking of hurricanes coming in from the coast of Africa. METEOSAT 3 started to move west over the Atlantic in 1989, and is currently positioned at 75° west longitude, giving better coverage of North America and the Atlantic. Interestingly, this satellite is now too close to the horizon for the European users to receive signals!

METEOSAT 3 has been a wonderful bonus for direct-readout users in North America, particularly those on the East Coast! But how does METEOSAT compare to GOES in the way of image products? Table 3 shows the main characteristics of METEOSAT series satellites:

Similar to GOES, METEOSAT spins the entire platform at 100 rpm. The main difference is that it spins from east to west, and scans south to north (unlike the GOES west to east spin, and north to south scan). Each pixel has a final resolution of around 2.5 kilometers.

The scanning method is as follows:

Method: Spinning platform/rotating instrument front end
Rate: 1.67 revolutions per second
Scan time: 600 milliseconds
Time for one line: 30 seconds

There are several data formats for the METEOSAT satellites. The "A" format represents the whole Earth-disc. The "B" format represents the northern part of the disc, with Europe in the middle. "X" format represents the North and South American continents.

The radiometer on METEOSAT can explore the Earth in three channels or spectral bands:

Visible: 0.4-1.1 µM
IR: 10.5-12.5 µM
Water vapor: 5.7-7.1 µM

The IR and Water Vapor are transmitted *inverted* relative to the GOES WEFAX (black is cold and white is warm).

Once again, several amateurs have come up with designs to receive the High Resolution (HR) imagery. (The low resolution WEFAX format [SDUS] may be received with exactly the same WEFAX equipment used for GOES, with the exception of the receiving frequency of 1694.5 MHz.) Marciano Righini and Guido Emiliani of Italy are two of the pioneers of direct readout imaging. They have developed a PDUS system that may be constructed from a kit. It was described in the British Remote Imaging Group journal in June 1993. Ed Murashie is working on a combined HRPT, VAS, and PDUS system that utilizes common circuit components. With his system you could receive three high-resolution image types on a single personal computer ground station! TimeStep, of Newmarket, England, currently markets a PDUS system for IBM personal computers. It is available for immediate ordering and installation (see references for address).

Figure 7.16 shows an example of METEOSAT 3 HR downlink of Southern Europe, Africa, and the Near East in a visible spectral band. This image was actually taken directly off the PC screen while doing real-time enhancement of the image with PC software from the NASA Jet Propulsion Laboratory.

NOAA SATELLITES: A BRIEF OVERVIEW OF CHANGES

In the Spring of 1996, a new series of operational environmental satellites will begin with the launch of NOAA-K (NOAA-15). NOAA-K, L, and M will be the successors to the current NOAA operated, polar orbiting satellites.

These new satellites will carry a series of instruments, which have been modified and improved from those now in orbit with the current operational satellites.

The Advanced Very High Resolution Radiometer (AVHRR/2) has been modified. The new instrument, AVHRR/3, adds a sixth channel in the near-IR, at 1.6 µM. This will be referred to as channel 3A and will operate during the daylight part of the orbit. Channel 3B corresponds to the previous channel 3 on the AVHRR/2 instrument, and will operate during the night portion of the orbit. The

Command :

Figure 7.16—A sample of METEOSAT 3 HR downlink imagery covering Southern Europe, Africa, and the Near East in the visible spectral band.

operational scheduling of the channel 3A/3B switching has not been precisely determined yet. A flag in word 22 of the telemetry will indicate which of the two channels is operating. Splitting channel 3 in this way maintains the HRPT data format, which was designed to handle five AVHRR channels. Channels 3A and 3B are output at the same telemetry locations.

Automatic Picture Transmission (APT) users will receive the AVHRR/3 channel 3A the same as channel 3B, with an ID wedge equivalent to gray scale wedge 3.

The AVHRR/3 visible channels (1, 2, and 3A) all have "split gains" or "dual slopes" that require the use of two calibration equations per channel, where previously one would suffice. The split gains, in effect, increase the sensitivity at low light levels.

The prime reason for these changes are to improve ice, snow and aerosol products produced from the visible channel data.

While NOAA-K will be tested in an afternoon orbit configuration, it will be capable of being launched as a morning or afternoon spacecraft to meet operational needs.

The Search and Rescue Processor (SAP) has added capabilities for the handling of distress messages, as well. The number of Data Recovery Units has been increased from two to three.

With the new instruments, data formats will be changed slightly. Within the HRPT Minor Frame format, the first Minor Frame will be TIP data, the second Minor Frame will be spare, and the third Minor Frame will be the AMSU-A1, -A2, and B data.

Within the TIP Minor Frame (orbital mode format), the deletion of the MSS and SSU instruments will make available several words. DCS/2 data will now be contained in additional words 18, 19, 24, 25, 32, 33, 40, 41, 44, 45, 61, 68, 69, 76, 77, 86 and 87. HIRS data will move from words 14/15 to 16/17. Word 102 will be a spare.

This information is provided to users of NOAA polar orbiting satellite data as an early indication of what changes they may expect with the launch of NOAA-K. It is preliminary in nature, however, and subject to revision prior to NOAA-K becoming operational in 1996.

GOES GEOSTATIONARY SATELLITES

GOES-8 and GOES-9 have now been launched with the new GVAR high-resolution imagery capable of 1 km resolution from 23,000 miles in space!

GOES-8 is at 75 degrees West longitude and is active with both the high resolution GVAR (replacement for the GOES SVAS data products) and low resolution WEFAX imagery.

GOES-9 is in "test" phase at 90 degrees West longitude and slowly being moved into position at 135 degrees West for its operational phase. GOES-9 also has high resolution GVAR and low resolution WEFAX data products available to the amateur enthusiast.

HRPT, VAS AND PDUS DATABASES

Many weather-satellite enthusiasts find it enjoyable to share the images they collect. It is always interesting to view HRPT images of Australia, Italy, Africa, and so on, and get a new viewpoint of the world. The Dallas Remote Imaging Group (DRIG) maintains a large database of APT, HRPT, WEFAX, VAS, and PDUS images on an electronic bulletin board. Users may dial in with their computer modems (any speed from 1200 bps to 16.8 kbps) and upload their images while downloading a collection of over 1000 weather-satellite pictures and images from other NASA, military, and amateur imaging spacecraft. The bulletin board system is operational 24 hours per day, and has access to INTERNET, FIDONET, NASA SPACELINK, and several on-line databases. Several hundred image processing programs, GIF viewers, satellite tracking programs, and satellite Keplerian elements are also available. Other electronic bulletin boards around the country carry some HRPT images that were collected and distributed by DRIG.

Weather satellite users are encouraged to join the Dallas Remote Imaging Group and become part of the Worldwide Satellite Tracking Network. This active group constantly monitors meteorological satellite frequencies and searches for new launches. Ground stations are set-up to automatically capture every pass and every image transmitted by meteorological satellites. Once a new satellite image is collected, electronic messages and faxes are sent worldwide to alert users to the new satellites. Users may contact the Dallas Remote Imaging Group by phone, fax, bulletin board, INTERNET, FIDONET, mail, and in person! (Users who connect with the DRIG BBS may actually send mail automatically over INTERNET and FIDONET e-mail networks worldwide!).

A free information kit is available for the asking. Please provide a 9 × 12 return envelope with $1.50 for postage. Write to:

Dallas Remote Imaging Group
Satellite Information Services
PO Box 117088
Carrollton, TX 75011-7088

BBS   214-394-7438 (24 hours)
Fax   214-492-7747 (24 hours)
Voice   214-394-7325
FIDONET   1:124/6509
CompuServe   76326,2447
Internet contact points:
  ftp.drig.com
  http://www.drig.com/
  BBS telnet to bbs.drig.com

DRIG also supports a monthly newsletter, the *Weather Satellite Report*, that contains the latest information on remote sensing and direct-readout systems. Refer to the reference at the end of this book for subscription information.

CONCLUSION

The demographics of the weather-satellite imaging community has changed drastically over the past ten years. What was once the private domain of a few technical amateurs, and large commercial ground stations, is now open to thousands of interested enthusiasts. High-resolution images from polar and geostationary orbiting satellites are available to be displayed on your personal computer on a daily basis. Images can be stored, processed, animated, and shared with others throughout the world. With a host of polar and geostationary spacecraft available to transmit meteorological images, there has never been a better time to upgrade your ground station and start enjoying the high-resolution View from Above!

Dr Wallach, N5ITU, is a private consultant in the weather satellite community. He has worked with NASA and NOAA for the past ten years to bring ground stations into classrooms to be used as a unique educational tool.

Chapter 8

Satellite Tracking

INTRODUCTION

In this chapter, you'll be introduced to the basic skills of satellite tracking. There's really nothing complicated about the process—you can do it all with a pencil, paper and elementary school math. Though the math is really quite simple, use of a pocket calculator helps to minimize the chance of error with repetitive calculations. In fact, there is no better subject than satellite orbits to introduce students to the tremendous predictive power of mathematics! Sure, most of the operations we'll cover here can be bypassed by the use of relatively simple computer programs. But if you jump right into tracking satellites by using a computer, you'll miss the chance to really understand the nature of satellite orbits. I'm firmly convinced that everyone can benefit from some experience with manual plotting and tracking techniques, and strongly urge you to acquire the skill—even if you intend to use your computer for day-to-day predictions.

ORBITAL-PLOTTING BOARD

One of the fundamental aids for any satellite station in determining satellite orbital tracks is a plotting board. Figures 8.1-8.3 provide the basic materials you'll need to construct such a board. You'll actually be making *two* plotting boards. One board you'll use in the sample exercises we'll step through; the other will be customized for your own station location. Here's how to create these two versions of the plotting board:

1) Make two photocopies of Figure 8.1. Glue each copy to a piece of poster board, or cardboard, to provide stiffening.

2) Make one copy of Figure 8.2. Use a pair of sharp scissors to carefully cut out the two circles, being careful to retain the N and S labels.

3) If you live in the northern hemisphere, make one copy of Figure 8.3, and use typist's opaquing (correction) fluid (or white paint) to cover up the numbers 51 to 101, adjacent to the points on the track. If you live in the southern hemisphere, make two copies of

Figure 8.3, cover the numbers 51 to 101 on one copy, and 0 to 50 on the second copy.

4) Take your copies of Figure 8.3 and have additional copies made on overhead-transparency film. (Most copy centers have this capability.) If you live in the northern hemisphere, a single copy is adequate. Make transparent-film copies of both of your copies of Figure 8.3, if you live in the southern hemisphere. If you don't have access to copy facilities that handle film material, use india ink or other indelible ink to trace the track arc, points, numbers, and the center cross on a moderately stiff piece of transparent acrylic or translucent Mylar sheet. There is no need to trace the circular or square border areas.

5) Use sharp scissors to cut out your film copy(s) around the circular margin, discarding the borders.

Now let's take a look at the pieces we have created and how they go together to make a plotting board. Figure 8.1 represents a polar projection of the earth, where the center point is either the north or south pole, depending on your location. The radial lines represent lines of longitude. The line coming straight down from the center point is 0°. Note that the longitude radii are labeled in 10° increments from 0 around to 350. Labeled in this manner, the longitude calibrations represent west longitude.

The concentric lines represent lines of latitude, beginning with the equator (0°) as the outermost circle, and working inward in 10° increments to the pole.

The diagram in Figure 8.2 represents azimuth bearings (N, NE, E, SE, S, SW, W, and NW), and elevation circles in 10° increments from 0° at the outermost circle, to 70°, with the center point representing 90°. Attach this diagram to the plotting board at our station location. For our sample tracking exercise, we'll deal with a hypothetical station at 80° W longitude and 40° N longitude. Take one of your copies of Figure 8.1 and use the point of a straight pin to mark this point on the diagram. Coat the back of one of your diagrams of Figure 8.2 with contact cement or other glue and insert the straight pin through the center of the diagram. Insert the point of the pin into the 80° W/40° N

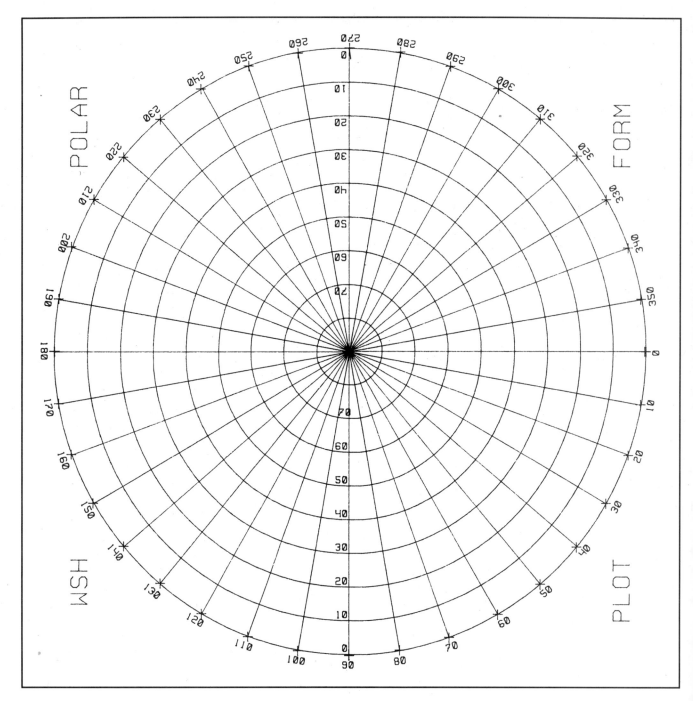

Figure 8.1—Polar plotting diagram.

point you previously marked. Slide the bearing circle down the pin, orienting it so that the N line points directly toward the center of the diagram. Smooth the bearing diagram into place. Your plotting board should now look like Figure 8.4, disregarding the plotting track.

Repeat these steps using your own station location on your second copy of Figure 8.1. If you are located in the southern hemisphere, consider the concentric latitude lines to represent south latitude, and the center point is the south pole. When placing the bearing diagram for a southern-hemisphere station, orient the diagram so that N points toward the pole. When the station diagram overlay has dried, relabel N on the diagram as S and S on the diagram as N.

To avoid smudging and soiling your own plotting board, cover the entire diagram (complete with the bearing circle at your station location) with clear

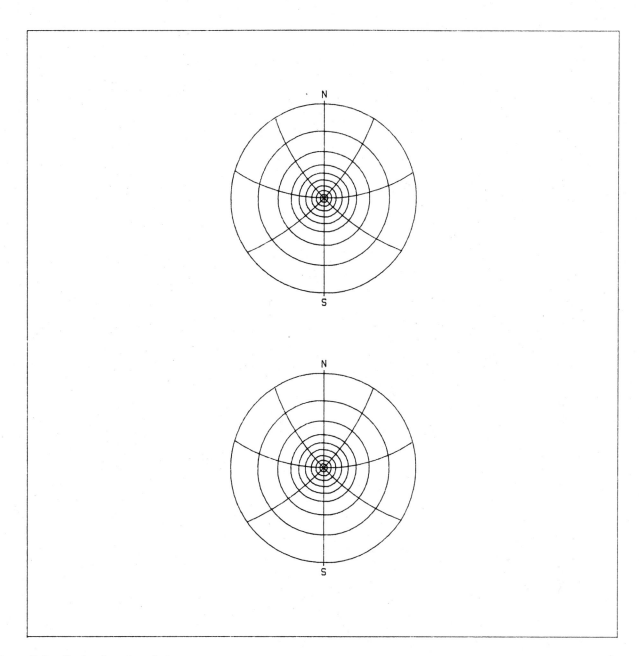

Figure 8.2—Station-bearing circles.

plastic film. There is no need to do this with the sample plot for our station at 80° W/40° N because we'll only use this version for a few sample exercises.

You should have one film copy of the track from Figure 8.3, with the track points labeled 0 to 50. Place a thumbtack through the center-cross of the film track, and insert the tack into the center of the diagram you prepared for our hypothetical 80° W/40° N station. You should now be able to rotate the track diagram around the pole. Set the diagram aside until we're ready for a sample tracking exercise. Once our sample exercises are complete, northern-hemisphere stations will transfer the track film to their own plotting board. Trackers in the southern hemisphere should attach the second copy of the track diagram (with points labeled 51 to 101) to their southern-hemisphere plotting board.

SATELLITE ORBITS

A satellite orbit is the path traced by the spacecraft in its trip around the earth. There are three major components that define the nature of this orbit: its eccentricity, period, and plane. *Eccentricity* defines the degree of departure of the orbit from circularity. Virtually all natural-object and satellite orbits are elliptical to some degree. In the case of an elliptical orbit, there is a point along the orbital path where the

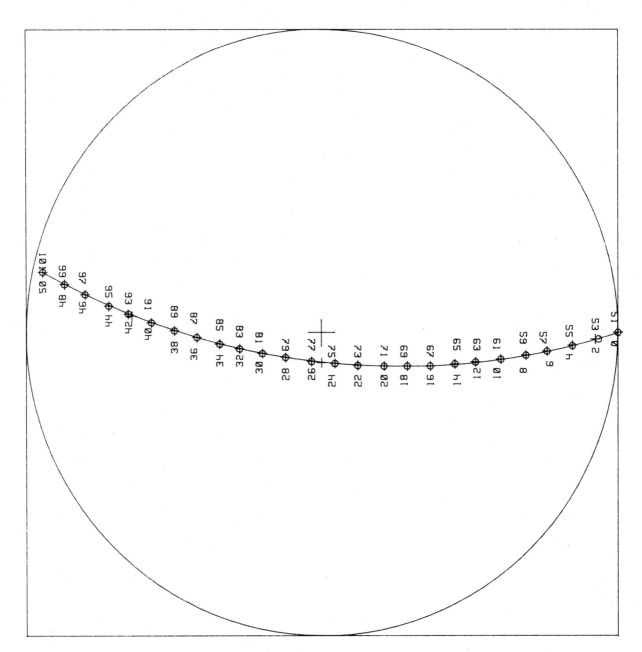

Figure 8.3—Reference orbital track.

satellite is closest to the earth (known as the *perigee*), while 180 degrees around the path is the *apogee*, or the point at which the satellite is farthest from the earth. For an operational weather-satellite satellite system, it is highly desirable to minimize the eccentricity (the ratio of apogee to perigee) and achieve something as close as possible to a circular orbit (apogee = perigee, eccentricity = 1). Although there is no such thing in nature as a perfectly circular orbit, injection of the TIROS/NOAA spacecraft into orbit is controlled as precisely as possible to yield orbits that are circular to within ±18.5 km during the operational lifetime of a given satellite.

The *period* of a satellite orbit is the time, in minutes, required for the satellite to complete one circuit of the earth. The period of the orbit is a function of the satellite's altitude. For near-circular orbits, the relationship can be expressed by:

$$P_{min} = 84.4 \times (1 + (h/r))^{3/2}$$

where P is the period (in minutes), h is the height, or altitude, of the orbit and r is the radius of the earth. As altitude increases, so does the period. The TIROS/NOAA satellites have an average altitude of about 854 km (460 nautical miles [nmi]), resulting in a period of 102 minutes. Geostationary satellites (such as GOES) orbit at an altitude of 35,790 km (19,312 nmi), resulting in a period of 1440 minutes or 24 hours. As a more extreme case, consider the earth, orbiting 93,000,000 miles "above" the sun. The earth follows precisely the same physical laws and requires approximately 365 days to complete an orbit as a satellite of the sun.

The *plane* of the orbit describes the relationship of the plane of the orbital path relative to the earth. An *equatorial* orbit is one in which the plane of the orbit lies in the plane of the equator. (This means that the orbital track lies directly over the equator.) Equatorial orbits are characteristic of the geostationary satellites such as GOES, METEOSAT, and GMS. Recall that these satellites have a period of precisely 24 hours (1440 minutes). Because the direction of their movement is the same as the direction of the earth's rotation beneath them, the satellites remain over the equator at the same point at all times—hence the term *geostationary*.

The mission of weather satellites in lower orbits is to photograph as much of the earth's surface as possible in the course of a day. Equatorial orbits don't achieve this goal at lower altitudes because so much of the earth north and south of the equator is out of view of the satellite. The ideal plane of a weather-satellite orbit is *polar*, where the plane is oriented 90° to the equator, such that the orbital track crosses the poles. As the earth rotates beneath the satellite, virtually all parts of the earth can be photographed during a day with the proper combination of altitude and period.

Precise polar orbits are not desirable if we want the satellite to pass overhead at roughly the same time each day. To achieve this, we require what is known as a *sun-synchronous* orbit. Power and thermal constraints rule out an orbit that produces passes within 2 hours of local noon (or midnight). The result is that TIROS satellite orbits have an inclination of 98.8° to the equator. This results in an orbital track that reaches a latitude of 81.2° (180 – 98.2)—what we refer to as a *near-polar* orbit.

If we're to predict where a satellite will be at a given time, we need some means to graphically present the orbit. At any point in its orbit, there is some point on the earth's surface that is directly beneath the satellite. This point is known as the satellite *subpoint*. If we plot the position of the subpoint at regular intervals during an orbit, we end up with an orbital track projected on to the earth's surface.

To deal with this orbital track in a predictive way, we need a place to start. The standard way to define the "start" of an orbit is to use the point at which the satellite crosses the equator in a north-bound direction. For any given satellite, this equatorial crossing occurs at a specific longitude and time on any given day.

PREDICT-DATA FORMAT

In order to predict where the satellite will be, we need data on a *reference crossing* for the satellite of interest. One of the most successful modes for disseminating this information over the years has been in the form of "Predict" post cards, mailed out each month by the satellite service. These cards (see Chapter 9 for more information) contain data on a single reference crossing (usually for the first orbit on the first day of the month) for each of the operational TIROS/NOAA satellites. An example, the following data was distributed for NOAA-10 for 01 December 1989:

	NOAA-10
ORBIT #	16642
EQ. XING TIME(Z)	0124.96Z
LONG. ASC. NODE(DEG)	88.89W
NODAL PERIOD	101.2340
FREQ. (MHZ)	137.50
INCR. BTWN ORBITS	25.31

ORBIT # represents the number of the reference orbit where orbit numbers are tabulated from the first orbit at launch.

EQ. XING TIME(Z) is the time when the crossing of the equator occurs. The NOAA postcard format treats the time format as follows:

HHMM.MM

where H is hours and M represents minutes to two decimal places. As you can see, orbit 16642 begins at 01 hours and 24.96 minutes.

As in all aspects of satellite work, the time is referenced to that at the prime meridian (0° longitude). The Z stands for Zulu time, a commonly used designator. Greenwich Mean Time (GMT) is another common (though obsolete) designation for prime-meridian time. The proper term to use is UTC. (Because of governmental use of Z, that letter and UTC will be used interchangeably here.) Although you can convert UTC to your local time, such conversions simply serve to increase the chance for error in your orbital predictions. Virtually all satellite stations maintain at least one clock set to UTC. The universal availability of inexpensive digital clocks makes this quite convenient, and provides a degree of accuracy more than sufficient for our purposes. You'll soon be adept at making conversions in your head in so far as knowing when the satellite's timetable interacts with your own personal schedule.

If you want to convert the decimal minutes value to seconds, simply multiply the decimal component (0.96 in the case of our NOAA reference data) by 60. Thus, in terms of hours, minutes, and seconds, our reference orbit (16642) begins at 01:24:58Z.

LONG. ASC NODE (DEG) refers to the longitude of the ascending node (in degrees); a way of stating the point (in degrees West longitude) where the satellite crosses the equator northbound at the start of orbit 16642 at 01:24:58Z—in this case, 88.86° W.

PERIOD represents the time (in minutes, to four decimal places) required for the satellite to complete each orbit. In the case of NOAA-10, each new orbit begins 101.234 minutes after the start of the last orbit. There are many factors that can effect the value of the period. Because the TIROS/NOAA satellites are at relatively low altitudes, residual atmospheric drag is a factor (which varies under the influence of many factors, including solar activity) that tend, over time, to slow the satellites, resulting in a slightly lower orbit and lowering the period. These variables are the primary limitation on how accurate we can be in simple, long-term predictions—a factor I'll discuss at greater length later in this chapter.

The FREQ. entry is simple: It tells us what transmitting frequency the satellite is using.

INCR. BTWN ORBITS is a bit obscure. It refers to the longitude increment between successive orbits. If the earth were not rotating, by the time the satellite had completed one orbit and was ready to start another, the new crossing would start at precisely the same subpoint on the equator as did the previous orbit. But, the earth *is* rotating at a rate of 360° every 24 hours (1440 minutes), or 0.25° per minute to the east. In the course of the 101.234-minute period required for one NOAA-10 orbit, the earth has rotated beneath the satellite a total of 101.234 × 0.25, or 25.3085 degrees toward the east. Because, within the course of a day, the satellite orbital plane is fixed, the next crossing point will be 25.3085 degrees to the *west* of the previous crossing. It is this value, rounded to 25.31, that represents the longitudinal increment between successive crossings. Because long-term predictions involve adding the increment repeatedly with each orbit, it is important that the increment be as accurate as possible. For this reason, I prefer to use the more accurate value derived from the period (as shown above) in preference to the rounded value included on the predict card.

With the combination of data provided on the Predict card and the concepts presented in the previous paragraphs, we have all the tools in hand to track the satellite in time—the subject of the next section.

PREDICTING CROSSINGS

Given reference crossing data of the type we have for NOAA-10, predicting each subsequent crossing is a simple matter:

- Add the period to the previous crossing time to get the new crossing time.

- Add the increment between orbits to the old crossing point to get the new crossing point.
- Increment the orbit number.

Before we actually go ahead and do this, there is one useful shortcut we can take that makes the job easier. Although we can add hours, minutes, and seconds for the time, the process is error-prone because calculators are not designed for the job. Things are much easier if we convert our reference crossing time to minutes:

$$T_{min} = (H \times 60) + M$$

Because our reference time for orbit 16642 is 01 hours and 24.96 minutes:

$$T = (1 \times 60) + 24.96$$
$$= 84.96 \text{ min}$$

Now we can add the period for each new orbit, keeping the time in the form of total minutes. We'll only bother to convert the time back to the HH:MM:SS format when needed—something we won't have to do with every orbit, as you'll see.

Next, take a piece of lined paper and make three columns, labeled ORBIT, TIME, and CROSSING. In the ORBIT column, enter 16642 to 16656 in sequence. The reason for stopping at 16656 will be evident shortly.

Now, load the value for period (101.2340) into your calculator's memory. This is usually done by keying in the number and pressing the M+ key on most calculators. Next, key in our starting time (84.96), and keep adding the value in memory (using the RM key) to the running total, recording the value in each succeeding TIME column until you reach ORBIT 16656. Your tally sheet should look something like this at the moment:

Orbit	Time (min.)	Crossing (°W)
16642	0084.960	
16643	0186.194	
16644	0287.428	
16645	0388.662	
16646	0489.896	
16647	0591.130	
16648	0692.364	
16649	0793.598	
16650	0894.832	
16651	0996.066	
16652	1097.300	
16653	1198.534	
16654	1299.768	
16655	1401.002	
16656	1502.236 (0062.236)	

Note that for orbit 16656, the TIME has reached a total of 1502.236 minutes. Because there are only 1440 minutes in a day, this orbit must occur during the next day—December 2. We can correct this time by subtracting 1440, yielding 62.236 for the first orbit on December 2. Assuming our starting orbit of 16642 is the first orbit of the day (it is, because if we subtract the period from the time, we would have a negative number), orbits 16642 through 16655 represent all possible crossings for the 1st of December and we already have the crossing times calculated.

Now we must calculate the crossing point for each orbit. Load the value for the increment between orbits (25.3085) into the calculator memory, key in the crossing point for orbit 16642 (88.86), and start a running total recording the values in the CROSSING column (rounded to the nearest whole degree) until you reach orbit 16653. Note that the total on your calculator is 367.2535. Because there are only 360 degrees in a circle, we have obviously worked our way *around the earth* in our subsequent crossings, until we are just past the prime meridian.

To correct for this, subtract 360 (yielding a value of 7.2535), and continue your calculations and recording of data. By the time you have reached orbit 16656, your sheet should look like this:

NOAA-10 01 December 1989

Orbit	Time (min)	Crossing (°W)
16642	0084.960	089
16643	0186.194	114
16644	0287.428	139
16645	0388.662	165
16646	0489.896	190
16647	0591.130	215
16648	0692.364	241
16649	0793.598	266
16650	0894.832	291
16651	0996.066	317
16652	1097.300	342
16653	1198.534	007
16654	1299.768	033
16655	1401.002	058
16656	0062.236	83

We now have the crossing times and crossing point for all of the orbits on December 1, as well as the first orbit (16656) on December 2. In the next section, we'll look at how to use our plotting board to convert these numbers into real tracking data!

WHERE IS THE SATELLITE?

Now that we know when and where the satellite will make equatorial crossings, we need the plotting board to see what these numbers mean in terms of the satellite track. If you spend much time listening to a specific satellite frequency, such as the 137.50-MHz frequency of NOAA-10, you already know that most of the time you'll not hear the satellite's signal. The reason for this will become clearly evident.

Take the plotting board you prepared for our sample 80° W/40° N station location. Let's look at the satellite track for the first orbit (16642) on our table. Rotate the overlay so the point label 0 is located at 89° W on the polar diagram. Your diagram should look like the one in Figure 8.4.

What your diagram is showing you is the path of the satellite for the first half of orbit 16642, with the subpoint noted every two minutes. The outermost bearing circle around our station location represents 0° elevation—our radio horizon. Unless the satellite track intersects this outer circle, we have no chance of hearing the satellite. Note that the track does intersect the outermost bearing circle, so we'll hear the satellite during this pass.

Because we might actually want to listen on this pass, the first step is to convert the crossing time in minutes back to an hour/minute/seconds format. This is accomplished in three steps:

1) HOURs are equal to the whole number obtained when we divide the total minutes by 60:

$$\text{HOURS} = \text{INT}(84.96/60) = \text{INT}(1.416) = 01$$

2) MINUTES are calculated by multiplying the remainder in Step 1 (0.416) by 60:

$$\text{MINUTES} = \text{INT}(0.416 \times 60) = 24.96 = 24$$

3) SECONDS are calculated by multiplying the remainder in Step 2 (0.96) by 60:

$$\text{SECONDS} = 0.96 \times 60 = 57.6 = 58$$

Thus, by our UTC clock, the crossing can be expected at 01:24:58Z. If we were using the Zapper omnidirectional antenna, we would expect acquisition of signal (AOS) from the satellite at about 5 minutes after the crossing (AOS = 01:29:58). The satellite will pass off to the west of our station (reaching a maximum elevation of slightly more than 15°, and we should get loss of signal (LOS) at about 17 minutes after the crossing (LOS = 01:41:58). This is not a particularly good pass, but if we wanted to record it, we could simply set a digital appliance timer to turn the recorder on at AOS (01:29) and off at LOS (01:42), and the pass would be recorded on tape for later playback.

Note that during orbit 16642, the satellite will approach from the south and disappear to the north. Such a track is known as an *ascending* pass.

If we were using a gain antenna and wanted to track the satellite, our plotting board could provide the

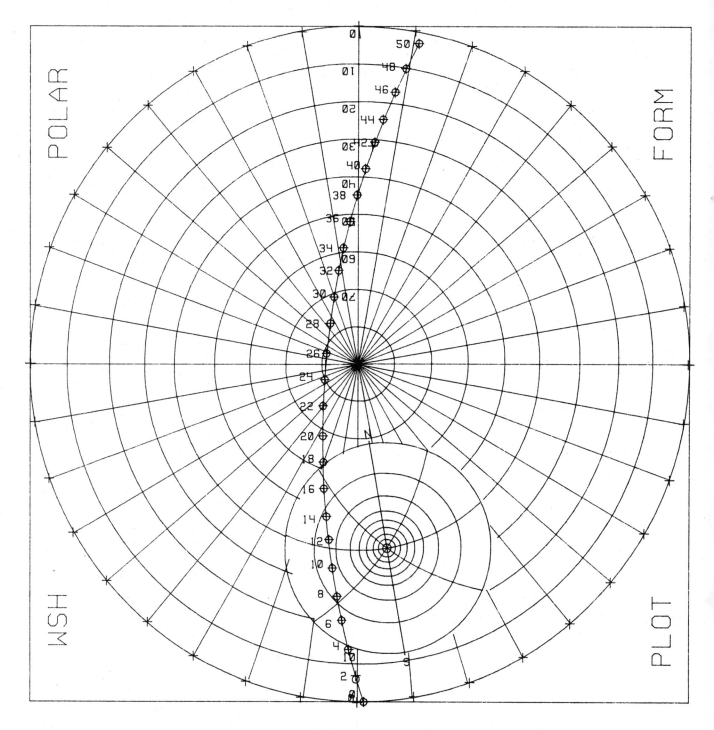

Figure 8.4—Sample plotting board layout for a reference crossing at 88° W.

basic information needed with a few more steps. Prepare a lined sheet with four labeled columns—MAC, TIME, ELEVATION, and AZIMUTH. In the MAC column (minutes after crossing), enter values from 5 to 17 at one-minute intervals. In the TIME column, compute the time equivalent for MAC = 5 by adding 5 minutes to the equatorial crossing (01:24:58 + 5 min

= 01:29:58. Fill in the remaining columns through MAC = 17 by adding one minute for each entry. The TIME for MAC = 17 should equal 01:41:58.

Now, using the plotted track (interpolating between the two-minute points), determine the elevation and azimuth at each one-minute interval. The concentric circles in Figure 8.4 represent elevation in 10° intervals from 0° to +70°. The radial lines represent azimuth (direction). When complete, your table should look something like this:

NOAA-10 Orbit 16642

MAC	Time	Elevation	Azimuth
5	01:29:58	4	SW
6	01:30:58	6	SW
7	01:31:58	9	SW
8	01:32:58	11	WSW
9	01:33:58	15	WSW
10	01:34:58	16	W
11	01:35:58	17	W
12	01:36:58	16	WNW
13	01:37:58	13	WNW
14	01:38:58	10	NW
15	01:39:58	7	NW
16	01:40:58	5	NW
17	01:41:58	2	NNW

If you were to set the antenna to the indicated elevation and azimuth values at each one-minute interval (either manually or with an az-el rotator system), you would successfully track the satellite during the time it was above the horizon.

A note here is appropriate for real purists. Our bearing diagram that is placed at the station location on our plotting board is simplified to a considerable degree. Although the elevation circles are plotted as circles, they have shapes that vary with station latitude. Similarly, the azimuth bearing lines are shallow arcs whose shape varies with station latitude. In the case of our bearing circle, the azimuth arcs are correct for 40° latitude. If we were using a very-high-gain/narrow-beamwidth antenna, the errors resulting from such simplifications could be significant. In the case of small 4- and 5-element beams, however, our simple, general-purpose bearing diagram is more than accurate enough to maintain a noise-free signal. If extreme tracking accuracy were required, in moving a small dish to acquire the HRPT signal for example, it would actually be easier to use one of the more elaborate tracking programs of the type I'll discuss later in this chapter.

Now, refer to your worksheet showing the other crossings for December 1, and we can proceed to check out the other orbits for the day. Rotate the overlay so the 0 point on the track is aligned with our next crossing point (114° for orbit 16643). Note that this pass is further west of our hypothetical station plotting and that it doesn't intersect the outermost elevation circle at all. We wouldn't hear the satellite at all during this pass at any point. If you proceed with orbits 16644 through 16647, you'll see that all are out of range of our station and thus of no interest. It

should be obvious now why we didn't bother with converting all our total minutes times. The only ones we need bother with are the ones associated with passes that bring the satellite above our local horizon.

If you check out orbit 16648 (241°), you'll see that it intersects the outermost elevation circle to the east of the station, but the maximum elevation is very low. Skip on to orbit 16649, and your plotting board should look like that of Figure 8.5. Make up a tracking worksheet like you did for orbit 16642, and attempt to calculate the times and antenna bearings as you did for 16642. Note, however, some crucial differences as you proceed. First, this is a *descending* pass—it originates to the north and you'll lose the satellite as it moves to the south, as opposed to the *ascending* mode illustrated by 16642. Also, pay close attention to the MAC times on the track plot. In this case, AOS occurs about 31 minutes after the reference crossing, with LOS at about 46 minutes after the crossing. If you have the idea, your worksheet should look something like this:

NOAA-10 Orbit 16649

MAC	Time	Elevation	Azimuth
31	13:44:36	0	N
32	13:45:36	4	N
33	13:46:36	8	N
34	13:47:36	11	N
35	13:48:36	17	N
36	13:49:36	24	NNW
37	13:50:36	30	NNW
38	13:51:36	40	NW
39	13:52:36	42	WNW
40	13:53:36	36	W
41	13:54:36	28	WSW
42	13:55:36	20	WSW
43	13:56:36	15	SW
44	13:57:36	9	SW
45	13:58:36	4	SW
46	13:59:36	0	SW

Note that this is the best pass we have seen so far (it is, in fact, the best of the day) because we have the satellite within range for about 15 minutes and it will reach a maximum elevation of about 42°.

After checking out the remaining passes for December 1, you are ready to put aside your sample plotting board and use the one that is set up for your station location. Repeat our tracking exercises for your station location and you are well on your way toward proficiency with the plotting board! Don't discard the sample board yet, however, for there is one more exercise we'll be doing with it after a few more skills have been added to your repertoire.

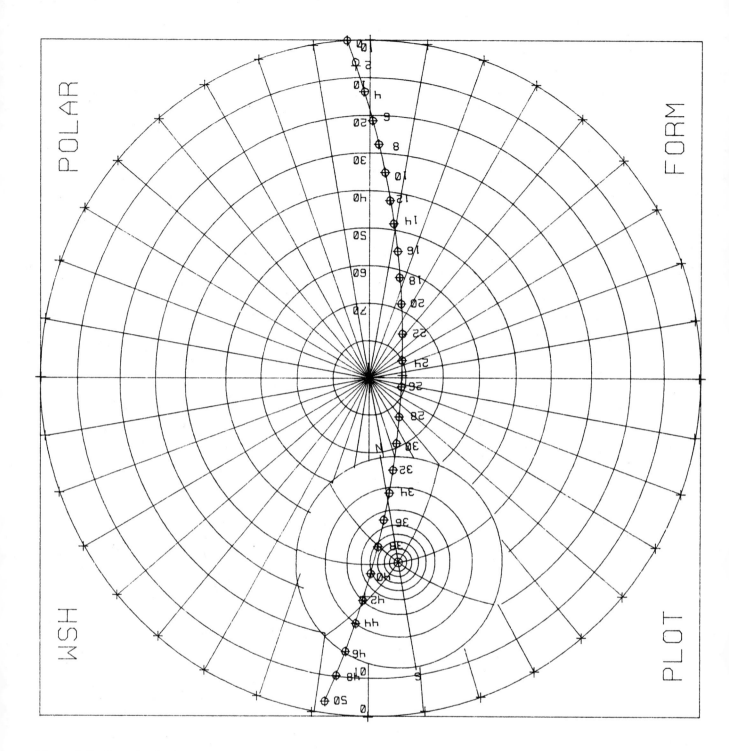

Figure 8.5—Sample plotting-board layout for a reference crossing at 266° W.

SOUTHERN-HEMISPHERE SATELLITE TRACKING

All our exercises so far have been appropriate for northern-hemisphere stations where we are concerned with the *first half* of each orbit. Southern-hemisphere stations, by contrast, are interested in the *second half* of the orbit (minutes 51 to 102). If you followed my earlier instructions on plotting-board preparation, your southern-hemisphere board will have a track overlay with subpoints labeled 51 to 101. Your tracking exercises involve one additional step.

Begin by setting the point on the overlay marked 51 at the crossing point for the orbit in question. This defines the ground track for the satellite during the first half (northern hemisphere) of the orbit. Note the point where the track intersects the equator at the far end of the track (just beyond 101). Mark this point, and rotate the overlay until point 51 is at that point on the equator. Now we have the track in place, properly labeled, for the second half (southern hemisphere) of the orbit and you can evaluate the pass relative to your location.

LONGER-TERM PREDICTIONS

If we were going to move through the month a day at a time, we could simply repeat what we have done previously. That's because the computation of each day's passes automatically yields the first orbit of the next day, which we can use as a starting point. If we had predict data for December 1, and we needed to track the satellite for the 15th, we could go through all the intervening days, but it would waste quite a bit of time. What we really need is a shortcut for jumping ahead by 14 days (15 − 1). The shortcut involves a series of 9 calculations:

1) First, determine the total number of minutes represented by the 14-day differential:

$$1440 \times 14 = 20160 \text{ min}$$

2) Divide the result obtained in Step 1 by the orbital period to determine how many orbits occur in that interval:

$$20160/101.2340 = 199.1426$$

3) To be sure that we'll land in the target day, we will use the next-highest whole number for total orbits: 200.

4) Determine the total number of minutes in 200 orbits:

$$200 \times 101.234 = 20246.8$$

5) Subtract the total obtained in Step 1 from the value in Step 4 to determine how much time must be added to our reference orbit:

$$20246.8 - 20160 = 86.8 \text{ min}$$

6) Add the result obtained in Step 5 to our reference time for orbit 16642 to get the orbital crossing time:

$$84.96 + 86.8 = 171.76 \text{ (new reference time)}$$

7) Add the result obtained in Step 3 to get the new reference-orbit number:

$$16642 + 200 = 16842$$

8) We must convert the time differential in Step 5 to a crossing point differential by multiplying by 0.25 (earth rotation/minute):

$$86.8 \times 0.25 = 21.7 \text{ degrees}$$

9) Add the result obtained in Step 8 to the original crossing point to get the new reference crossing point:

$$88.86 + 21.7 = 110.56 \text{ degrees}$$

In just 9 steps, as opposed to the dreary calculation of 200 intervening orbits, we have a reference crossing for December 15:

Orbit	Time	Crossing
16842	171.76Z	110.56W

Note that because the time is 171.76, while the period equals 101.234 minutes, there is actually one earlier orbit on this particular day. At most locations this may be irrelevant, but you can calculate the values for this earlier orbit by decrementing the orbit number by 1, subtracting the period from the crossing time, and subtracting the interval between orbits (25.3085) from the crossing point. At this point, the remaining orbits for December 15 can be handled just as we did for December 1.

THE CONCEPT OF PASS WINDOWS

It should be obvious at this point that only a small number of orbits each day yield useful passes at any specific station location. It turns out to be quite easy to predict which passes these will be, based on the equatorial crossing point. The best possible passes are the ones where the satellite passes directly overhead. For either an ascending or descending pass, this can only happen if the crossing occurs as a specific point.

Using the sample plotting board, rotate the overlay until the satellite track passes directly over the center point for the sample station location in the ascending mode. If you now examine the location of the 0 subpoint, it should be located at about 69° W longitude. For this hypothetical station location, an ascending overhead pass can *only* occur with a crossing at 69°. While the overlay is in the overhead position, note the times when the satellite will be overhead. AOS will occur at about four minutes after the crossing, the satellite will be overhead approximately 11 minutes after the crossing, and LOS will occur at about 19 minutes.

Although overhead passes are relatively rare, each day you'll have one pass that comes closest to the overhead orientation. Because the increment between orbits is 25.3085 degrees (a value we'll round to 26° for convenience), this "best ascending pass" must fall within ±13 degrees of the nominal 69° overhead crossing point. To be a bit conservative, we could widen this window to ±15°. If we are interested in the best possible ascending pass of the day for *any* satellite in the TIROS/NOAA series, only passes within the window of 54° (69 − 15) and 84° (69 + 15) need concern us. At least one pass must fall within this window during the course of a day. In the case of our December 1 data for NOAA-10, orbit 16655 is the best of the lot for an ascending pass. As far as times are concerned, an AOS of +4 minutes, closest approach to overhead at +11 minutes, and LOS at +19 minutes, derived from the overhead track, are essentially accurate enough for any pass within the best-ascending-pass window.

A similar exercise can be performed for the descending configuration by orienting the overlay to produce a descending overhead pass. This should occur with a crossing of 257°. In this case, AOS occurs at about +32 minutes, overhead at about +39 and LOS at +47. Our best descending window must thus fall between 242° (257 − 15) and 272° (257 + 15).

Summarizing this information for our hypothetical station, we get a short table that looks something like this:

	ASCENDING	DESCENDING
Min. Crossing	54	242
Overhead Crossing	69	257
Max. Crossing	84	272
AOS time	+4	+32
Overhead time	+11	+39
LOS time	+19	+47

If you use an omnidirectional antenna and a timer connected to the station tape recorder, there's no need to consult the plotting board at all to determine which passes to record. To select the best ascending pass of the day, pick the best pass (typically there's only one) that falls in the 54- to 84-degree window, and simply set the timer to go on four minutes before the crossing and turn off at 19 minutes after the crossing. For descending passes, simply scan the calculations for the pass within the 242- to 272-degree window, set the timer for ON at +32 minutes and OFF at +47 minutes, relative to the crossing time.

If you've followed the discussion this far, you can proceed to construct a similar table for your own station location. Both your table and the sample given previously are needed in the discussion of the computer program that follows in the next section.

A SIMPLE PREDICT PROGRAM

It should be obvious that the major problem with manual tracking is the inordinate number of calculations that have to be made, even with a memory calculator. Much of this burden is eliminated by the use of a simple computer tracking program such as the one in Figure 8.6. The program is written in GW BASIC and runs without modification on IBM and compatible PCs. Very little, if any, alteration of the program should be required to run under other BASIC dialects.

Entering the Program Listing

Accuracy is the key to successfully keying in the program. If you are generally unfamiliar with BASIC programming, be aware that small typos can generate large numbers of SYNTAX errors when you first attempt to run the program, and each must be fixed before the program will operate properly. Numerical entry errors may not generate error messages, but can lead to incorrect results. Take your time! As one time-saver, all the REM lines (REMarks) can be omitted on your copy of the program provided you retain the listing for reference as you try to understand what the program is doing. When you have the program entered, save a copy to disk or tape. You'll probably have some editing to do to correct errors, but the saved copy eliminates the need to re-do the job if a catastrophe occurs. As you gradually edit out the inevitable errors, keep saving the program until you finally have a clean master copy.

Language and Computer Compatibility

The program runs on IBM and other PC compatibles with no problem and should work well on most other systems with fairly advanced BASIC features. The only two problems you may run into on other systems concerns variable names and screen formats.

Advanced versions of BASIC allow long variable names (such as PASSWINDOW), and I have used such long names to make the variables as obvious as possible. Some 8-bit home-computer BASIC dialects support only two-character-long variables. If this is the case with your system, go through and assign rational abbreviations to the various variable names (such as PW for PASSWINDOW), being sure to catch all occurrences of each variable and array name. If you know in advance that you'll have this problem, make the changes as you enter the listing, keeping a running list of your abbreviations for reference as you move through the program.

The program is set up for an 80-character-wide screen display. If your computer cannot handle such a display, you'll have to rework the various PRINT statements to fit your available screen size.

There are no unusual functions used in this program, so other programming-language problems are not likely to arise.

Running the Program

The program, as listed, is customized for our hypothetical 80° W/40° N station location, and includes satellite reference data for December 1, 1989. If you first test the program with this data, you can cross-check the results against your manual calculations to verify that all is working well. We'll cover the business

PREDICT.BAS Program Listing

```
1Ø   CLS
2Ø   REM ********************************************************
3Ø   REM STATION DATA
4Ø   PASSWINDOW = 15
5Ø   ASCLONG = 69
6Ø   ASCOVERTIME = 11
7Ø   DESLONG = 257
8Ø   DESOVERTIME = 4Ø
9Ø   REM ********************************************************
1ØØ  REM READ SPACECRAFT DATA ON FILE
11Ø  READ NUMENTRIES
12Ø  READ DAY$
13Ø  FOR N = 1 TO NUMENTRIES
14Ø  READ SPACECRAFT$, ORBIT, TIME, CROSSING, PERIOD, FREQUENCY$
15Ø  SPACECRAFT$(N)=SPACECRAFT$:ORBIT(N)=ORBIT:TIME(N)=TIME
16Ø  CROSSING(N)=CROSSING:PERIOD(N)=PERIOD:FREQUENCY$(N)=FREQUENCY$
17Ø  NEXT N
18Ø  REM ********************************************************
19Ø  REM DISPLAY DATA ON FILE
2ØØ  REM PRINT BANNER
21Ø  GOSUB 159Ø
22Ø  PRINT:PRINT
23Ø  X$="FILE REFERENCE DATE: "+DAY$:GOSUB 171Ø
24Ø  PRINT
25Ø  FOR N = 1 TO NUMENTRIES
26Ø  N$ = STR$(N):N$ = RIGHT$(N$,1)
27Ø  X$ = "("+N$+")  "+SPACECRAFT$(N):GOSUB 171Ø
28Ø  NEXT N
29Ø  PRINT:PRINT
3ØØ  X$ = "KEY NUMBER FOR DESIRED SPACECRAFT OR Ø TO QUIT....":GOSUB 171Ø
31Ø  Q$ = INKEY$:IF Q$="" THEN GOTO 31Ø
32Ø  CLS:IF Q$ = "Ø" THEN END
33Ø  N = VAL(Q$):IF (N>NUMENTRIES)OR(N=Ø) THEN GOTO 19Ø
34Ø  REM ********************************************************
35Ø  REM SET REFERENCE DATA VALUES FOR TARGET SPACECRAFT
36Ø  SPACECRAFT$=SPACECRAFT$(N)
37Ø  ORBIT = ORBIT(N)
38Ø  TIME = TIME(N)
39Ø  CROSSING = CROSSING(N)
4ØØ  PERIOD = PERIOD(N)
41Ø  FREQUENCY$ = FREQUENCY$(N)
42Ø  REM ********************************************************
43Ø  REM CONVERT NOOA TIME FORMAT TO TOTAL MINUTES
44Ø  HOUR = INT(TIME/1ØØ)
45Ø  MINUTES = TIME - (HOUR*1ØØ)
46Ø  TIME = (HOUR * 6Ø) + MINUTES
47Ø  DAY = 1
48Ø  REM ********************************************************
49Ø  REM GET TARGET DAY
5ØØ  CLS
51Ø  GOSUB 159Ø:PRINT:PRINT:PRINT
52Ø  X$="REFERENCE DATE: "+DAY$:GOSUB 171Ø
53Ø  PRINT:PRINT
```

Figure 8.6—Listing for the WSH PREDICT.BAS program.

```
540 PRINT"DESIRED DAY FOR ";SPACECRAFT$;:INPUT TARGETDAY:CLS
550 REM ***************************************************
560 GOSUB 1590:PRINT:PRINT:PRINT
570 X$="COMPUTING CROSSING DATA":GOSUB 1710
580 REM CHECK IF TARGET DAY MATCHES CURRENT DAY
590 IF DAY = TARGETDAY THEN GOTO 620
600 GOSUB 1030:GOTO 590
610 REM ***************************************************
620 REM PRINT CROSSING DATA
630 CLS
640 REM PRINT BANNER
650 GOSUB 1590:PRINT
660 PRINT "SPACECRAFT:      ";SPACECRAFT$;TAB(50)"FREQUENCY: ";FREQUENCY$:PRINT
670 PRINT "REFERENCE DAY: ";DAY$;TAB(50)"CURRENT DAY:";TARGETDAY:PRINT
680 PRINT"------------------------------------------------------------------"
690 PRINT TAB(10);"CROSSING";TAB(20);"CROSSING";TAB(30);"AOS";
700 PRINT TAB(40);"LOS"
710 PRINT "ORBIT";TAB(10)"TIME (Z)";TAB(20);"DEG W";TAB(30);"TIME";
720 PRINT TAB(40);"TIME";TAB(50);"TRACK"
730 PRINT "------------------------------------------------------------------"
740 REM ***************************************************
750 REM TEST CROSSING AGAINST WINDOW AND PRINT IF FIT
760 GOSUB 1150
770 REM ***************************************************
780 REM COMPUTE NEXT ORBIT
790 GOSUB 1030
800 REM CHECK FOR NEXT DAY
810 IF DAY > TARGETDAY THEN GOTO 830
820 GOTO 750
830 PRINT "------------------------------------------------------------------"
840 X$="NEXT DAY (Y/N)?":GOSUB 1710
850 Q$ = INKEY$:IF Q$ = "" THEN GOTO 850
860 IF (Q$="y")OR(Q$="Y") THEN TARGETDAY = DAY:GOTO 620
870 IF (Q$="n")OR(Q$="N") THEN CLS:GOTO 190
880 GOTO 850
890 REM ***************************************************
900 REM CONVERT TIME IN MINUTES TO H:M:S STRING
910 HOUR = INT(TIME/60):X$=STR$(HOUR):GOSUB 980:HOUR$=X$
920 MINUTES = INT(TIME-(HOUR*60))
930 X$ = STR$(MINUTES):GOSUB 980:MINUTE$=X$
940 SECONDS = TIME -((HOUR*60)+MINUTES):SECONDS = SECONDS * 60
950 SECONDS = INT(SECONDS + .5):X$ = STR$(SECONDS):GOSUB 980:SECONDS$ = X$
960 CLOCK$ = HOUR$+":"+MINUTE$+":"+SECONDS$:RETURN
970 REM ***************************************************
980 REM FORMAT STRING COMPONENTS
990 IF LEN(X$)=3 THEN X$ = RIGHT$(X$,2):GOTO 1010
1000 X$=RIGHT$(X$,1):X$ = "0"+X$
1010 RETURN
1020 REM ***************************************************
1030 REM NEXT ORBIT ROUTINE
1040 REM UPDATE TIME
1050 TIME = TIME + PERIOD:IF TIME<1440 THEN GOTO 1090
1060 REM UPDATE DAY AND INCREMENT PRECESSION
1070 TIME = TIME - 1440:DAY = DAY +1:CROSSING = CROSSING +(360/365)
1080 REM UPDATE CROSSING POINT
```

Figure 8.6—Listing for the WSH PREDICT.BAS program (continued).

```
1090 CROSSING = CROSSING + ((PERIOD/1440)*360)
1100 IF CROSSING >= 360 THEN CROSSING = CROSSING - 360
1110 REM UPDATE ORBIT NUMBER
1120 ORBIT = ORBIT + 1
1130 RETURN
1140 REM ********************************************************
1150 REM WINDOW TEST
1160 C = INT(CROSSING + .5)
1170 REM SET WINDOW BOUNDARIES
1180 AMAX = ASCLONG + PASSWINDOW
1190 AMIN = ASCLONG - PASSWINDOW
1200 DMAX = DESLONG + PASSWINDOW
1210 DMIN = DESLONG - PASSWINDOW
1220 REM CHECK FOR ASCENDING WINDOW FIT
1230 IF (C<AMAX)AND(C>AMIN) THEN GOTO 1270
1240 REM CHECK FOR DESCENDING WINDOW FIT
1250 IF (C<DMAX)AND(C>DMIN) THEN GOTO 1310
1260 GOTO 1580
1270 REM ASCENDING WINDOW
1280 CENTER = ASCLONG
1290 TRACK1$ = "ASCENDING "
1300 GOTO 1350
1310 REM DESCENDING WINDOW
1320 CENTER = DESLONG
1330 TRACK1$ = "DESCENDING "
1340 GOTO 1350
1350 REM DETERMINE TRACK LABEL
1360 IF C = CENTER THEN TRACK2$ = "OVERHEAD":GOTO 1430
1370 IF C>CENTER THEN TRACK2$ = "WEST":GOTO 1410
1380 TRACK2$ = "EAST"
1390 IF C>=CENTER -5 THEN TRACK2$ = "NEAR OVERHEAD "+TRACK2$
1400 GOTO 1430
1410 IF C<=CENTER + 5 THEN TRACK2$ = "NEAR OVERHEAD "+TRACK2$
1420 GOTO 1430
1430 TRACK$ = TRACK1$ + TRACK2$
1440 REM COMPUTE CROSSING TIME STRING
1450 GOSUB 900: CTIME$ = CLOCK$
1460 REM COMPUTE AOS TIME STRING
1470 T = TIME
1480 IF TRACK1$ = "ASCENDING " THEN TIME = T + ASCOVERTIME:GOTO 1500
1490 TIME = T + DESOVERTIME
1500 OT = TIME
1510 TIME = OT-7:GOSUB 900:AOSTIME$ = LEFT$(CLOCK$,5)
1520 REM COMPUTE LOS TIME STRING
1530 TIME = OT+7:GOSUB 900:LOSTIME$ = LEFT$(CLOCK$,5)
1540 REM PRINT ORBITAL VALUES
1550 TIME = T
1560 PRINT ORBIT;TAB(10);CTIME$;TAB(20);C;TAB(30);AOSTIME$;
1570 PRINT TAB(40);LOSTIME$;TAB(50);TRACK$
1580 RETURN
1590 REM ********************************************************
1600 REM PRINT BANNER
1610 X$="*****************************":GOSUB 1710
1620 X$="* POLAR ORBIT PREDICT PROGRAM *":GOSUB 1710
1630 X$="*                             *":GOSUB 1710
```

Figure 8.6—Listing for the WSH PREDICT.BAS program (continued).

```
164Ø X$="*                VER. WSH4              *":GOSUB 171Ø
165Ø X$="*                                       *":GOSUB 171Ø
166Ø X$="*        DR. RALPH E. TAGGART           *":GOSUB 171Ø
167Ø X$="****************************************":GOSUB 171Ø
168Ø PRINT
169Ø RETURN
17ØØ REM *************************************************
171Ø REM FORMAT TO CENTER SCREEN DATA (8Ø COLUMN)
172Ø L = LEN(X$):L=INT(L/2)
173Ø LEADBLANKS = 4Ø-L
174Ø PRINT TAB(LEADBLANKS);X$
175Ø RETURN
18ØØ REM *************************************************
181Ø REM SPACECRAFT FILE DATA
182Ø REM NUMBER OF SPACECRAFT ON FILE
183Ø DATA 3
184Ø REM REFERENCE DATE
185Ø DATA "Ø1 DEC 1989"
186Ø REM DATA FORMAT
187Ø REM SPACECRAFT, ORBIT, TIME(Z), CROSSING(W), PERIOD, FREQUENCY
188Ø DATA "NOAA 9", 256Ø1, 13Ø.75, 127.14, 1Ø2.Ø213, "137.62 MHZ"
189Ø DATA "NOAA 1Ø", 16642, 124.96, 88.86, 1Ø1.234, "137.5Ø MHZ"
19ØØ DATA "NOAA 11",  61Ø1, 55.53, 165.32, 1Ø2.Ø851, "137.62 MHZ"
2ØØØ REM ************ END OF PROGRAM *************************
```

Figure 8.6—Listing for the WSH PREDICT.BAS program (continued).

of customizing for your location and entering new satellite data in the sections that follow.

When you first run the program, you'll be given the date for the satellite reference data on file and a list of the operational satellites. In this case, there will be three of them: NOAA-9, NOAA-10, and NOAA-11. Press one of the number keys (1 to 3) to select one of these satellites and get things rolling. If you press Ø, the program ends. Pressing other alphanumeric keys has no effect. For the sake of comparability with all the manual calculations you have already performed, press 2 for NOAA-10.

Once you have selected a satellite, you are asked to enter a day for the readout. Press 1 for December 1, and hit the ENTER or RETURN key. Within a second or so, the information on best passes for the day appears on your screen. The data provided includes the orbit number, the time of the crossing (in an HH:MM:SS format), the crossing point (rounded to the nearest whole degree), AOS and LOS times (in an HH:MM format), and a verbal description of the track. The track description tells you whether the pass is ascending (A) or descending (D), and the path of the track. Overhead passes are labeled. Passes labeled "near overhead east," or "near overhead west" are those that fall within 5° of the nominal overhead crossing points. All other passes are simply labeled "east" or "west," depending upon the orientation of the track relative to your station.

Once these data are displayed, you can move to the next day by pressing Y, or return to the satellite-selection menu by pressing N. Use Y to select the next few days to get a feel for how much time the program saves you, and then press N for the satellite menu.

Any day of December can be selected once you have chosen a satellite. Note that there is no direct provision for going into another month. This is because a simple program like this one accumulates errors resulting from the low orbits of the TIROS/NOAA satellites, and these errors begin to get significant as we stretch out toward the end of the month. You can get data for the next month by simply entering a number beyond the end of December. December has 31 days; should you enter 32, you'll get the display information for the first day of the next month—in this case, January 1. This extension feature is provided just to tide you over if your predict data are late.

Customizing the Program

Once you have the program working as listed, save a copy for reference, then proceed to edit lines 4Ø-8Ø to customize the program for your station location:

```
4Ø    PASSWINDOW = 15
5Ø    ASCLONG = 69
6Ø    ASCOVERTIME = 11
7Ø    DESLONG = 257
8Ø    DESOVERTIME = 39
```

All of these values should be familiar to you because they are the ones we derived for our hypothetical station when looking at the subject of pass windows. PASSWINDOW is the width, in degrees, for one-half of the pass window, centered on the nominal overhead crossing point. If you narrow this window, you'll miss some of the best passes. If you widen the window to about 20, you'll include a few additional passes. Do not widen much beyond this point, or you'll be dealing with marginal passes to the east or west.

ASCLONG is the longitude for an ascending overhead pass. Substitute the value calculated for your location. ASCOVERTIME is the nominal time to overhead following the crossing. Again, substitute your values. DESLONG and DESOVERTIME are the corresponding values for descending passes and you should substitute your values.

Now you can save the program as the master copy of your working version and proceed to check out passes in December for your location. If you have previously worked out a series of passes manually, you can compare the program results against this earlier work.

Updating Satellite Data

For ease of use with all computer systems, the reference crossing data are incorporated at the end of the program in the form of data statements. Line 1830 defines the number of satellites on file, while 1850 contains a string for the date of the reference crossing. Line 1880 contains the crossing data for NOAA-9, 1890 covers NOAA-10, and 1900 covers NOAA-11. The principle advantage of this approach is that the program will run on computers not equipped with disk drives. To use this feature, you'll have to edit these lines once each month when your new predict data arrives, inserting the new dates and satellite crossing data. Save the revised working copy for use throughout the month.

If your computer system has disk drives, as most now do, it is a simple matter to rewrite the file statements to load the data from a disk file. This file would have to be updated monthly, however, so it is questionable whether the disk version would be any more convenient.

Chapter 9 includes information on how to get on the NOAA mailing list for a Predict card, or, better yet, how to obtain the data from one of the many electronic bulletin-board systems.

The *WSH Program Disk*, available from the ARRL, was discussed in Chapter 7 in terms of the WSH1700 program for control of the scan converter. This program disk also contains a more-sophisticated tracking program for IBM PCs and compatibles. When you first run the program, data on your station location is requested and stored in a disk file. Another disk file is constructed based on information from your first Pre-

dict card. Because it's a compiled program, it runs very fast. The program itself calculates best-pass windows and other data that had to be entered manually into the simpler program presented here. While still simpler than the more-elaborate tracking programs, the disk-based program is very easy to use and is sufficient for stations using an omnidirectional antenna system.

Long-term Accuracy

If the satellite were in an absolute vacuum, orbiting around a perfectly spherical earth with absolutely uniform internal mass distribution (and there was nothing else going on in the universe), this simple program would produce results that could be used for a very long time after the reference crossing. Unfortunately, none of these conditions are true. The result is that the period of the satellite is constantly changing with a *general* trend toward a shortening of the period as the satellite's orbit gradually decays. It is these changes in the period that cause errors to build up in longer-term predictions because the period errors accumulate arithmetically with every orbit. If our orbital value is off by 0.001 minutes (less than a tenth of a second) by the end of a month, our crossing time can be in error by almost half a minute! If the error is in the order of 0.01 minutes, we'll be over four minutes off target by month's end! The major causes of such errors are the natural perturbations of the satellite's orbit and the precision of the initial orbital value that we start with in our program. We can't do much about the former, but we can attack the latter.

A graphic case of the impact of such errors arose when I was testing the program we have just discussed. It is a very simple program (the one I normally use has a listing that is over 14 pages long!), but there were errors by month's end that simply seemed too great. There are two ways to benchmark a program such as the one presented here. The first is by careful analysis of the imagery to determine the *actual* time the satellite subpoint reaches a known latitude, then backtrack to calculate what the crossing time must have been. The second approach, which I was using, was to compare the results generated by this simple program with a more-complex program of known accuracy.

There are a number of superb tracking programs available and one excellent one, SAT TRAK (written by William N. Barker and David G. Cooke and modified by T.S. Kelso), is available—along with many others—on the DRIG BBS, which I'll discuss in Chapter 9. This program uses Keplerian orbital elements (that are constantly updated and distributed by NASA) in conjunction with sophisticated algorithms that even take into account the elevation of your ground station.

The information presented by such programs is very precise (particularly if you are using the most recent set of orbital elements), and SAT TRAK can be used (in conjunction with a powerful telescope), to visually track satellites when lighting conditions are suitable. If you need precision tracking data, such a program is ideal, but it is more complex to use and (given the detail provided in the printouts) has a tendency to use up computer paper. PREDICT, on the other hand, is designed to be very easy to use, with the aim of generating primarily the data needed to make optimum use of an omnidirectional antenna system. In any case, when SAT TRAK was used to cross-check the results generated by PREDICT for December 26, the following results were obtained:

NOAA-9

	AOS	LOS
PREDICT (NOAA data)	10:46	11:00
SAT TRAK (NASA data)	10:44	10:58

NOAA-10

	AOS	LOS
PREDICT (NOAA data)	00:23	00:42
SAT TRAK (NASA data)	00:25	00:39

NOAA-11

	AOS	LOS
PREDICT (NOAA data)	18:37	18:51
SAT TRAK (NASA data)	18:38	18:52

Just for the record, the various satellites arrived right on schedule in terms of the SAT TRAK data. My problem was how to deal with the errors in PREDICT's listing. Results varied with individual satellites (to be expected), but the bottom line was that PREDICT was two minutes late with NOAA-9, one minute early with NOAA-10, and one minute early with NOAA-11. These errors weren't monumental, but they are great enough to mess up a tracking exercise. The crossing-point calculations were quite accurate. The NOAA-10 pass, for example, was printed out as an ascending overhead pass. In the SAT TRAK printout, maximum satellite elevation was 86.6°. Considering that PREDICT rounds crossing data to the nearest degree when evaluating the track, the results were quite acceptable!

Once math errors had been ruled out, the only possible variable was the value for the period. The following period values were contained in the NOAA Predict Bulletin for December 1, 1989:

NOAA-9	102.0213 min
NOAA-10	101.2340 min
NOAA-11	102.0851 min

I was immediately suspicious of the value for NOAA-10 because a check of past prediction bulletins showed that this value had not changed in several months. Several phone calls to the satellite service verified that

they had been using a new procedure for transferring period data from the voluminous NASA printouts they receive; the result was small, but significant, period errors. They have now adopted a different procedure that should minimize the problem in the future. Our problem, if we use the monthly bulletins, is how to determine if the period supplied is actually the best one for our use. It turns out that there are two relatively easy ways to derive a high-precision value for the period that can improve our calculations immensely.

The first scheme makes use of the two-line NASA Keplerian element sets that are available (and updated several times each month) on bulletin boards such as the DRIG system. The format for these element sets looks cryptic simply because an incredible amount of information is packed into a limited printout. An example of one of these element sets for NOAA-10 is shown below:

NOAA-10
1 16969U 86 73 A 89<u>353.40645465</u> .00000892 00000-0 40960-3 0 3183
2 16969 98.6208 20.2379 0014656 72.1749 288.1026 <u>14.23364909</u>169038

I have underlined an item in line 1, and another in line 2. These are not underlined in the actual data set, but I did this to make it easier to point out the two items of interest when discussing the PREDICT program. There are excellent tutorials available on the DRIG BBS for decoding all of the information in the element set but, for our purposes, we'll concentrate on the two underlined items.

The item on line 1 represents the Julian day for this particular data—the December 19. The preceding digits (89) represent the year. The underlined item on the second line tells us that NOAA-10 makes 14.23364909 orbits on that day. If we divide the number of minutes in a day (1440) by this value, we should end up with a period:

$$1440/14.23364909 = 101.16871583$$

While this *is* a period value, it represents what is known as the *anomalistic period*. To convert the anomalistic period to our more familiar nodal period, we multiply the anomalistic value by 1.0056:

$$101.16871583 \times 1.00056 = 101.22537031$$

Note that this value for the nodal period is different than the 101.2340 minutes in the original NOAA predict listing. If this "refined" value, as well as comparable values for the other two satellites, is plugged into the program in place of the NOAA values, we get the following:

	NOAA-9	
	AOS	LOS
PREDICT (refined data)	10:44	20:58
SAT TRAK (NASA data)	10:44	10:58

	NOAA-10	
	AOS	LOS
PREDICT (refined data)	00:24	00:38
SAT TRAK (NASA data)	00:25	00:39

	NOAA-11	
	AOS	LOS
PREDICT (refined data)	18:38	18:52
SAT TRAK (NASA data)	18:38	18:52

Note that most of the errors between the two sets of results have disappeared except for a one-minute offset on the NOAA-10 data. Given the fact that PREDICT rounds AOS time calculations to the nearest minute, a one-minute error is about the maximum that will be encountered. If you set your timer for one minute earlier and later than the calculated AOS and LOS times, you'll probably never miss anything of consequence.

There is a second source of more refined period data: That is the periodic NASA TBUS bulletins that are transmitted daily on world meteorological HF frequencies, on WEFAX, and which are available on some bulletin boards. TBUS messages are transmitted in four parts for each satellite. The one that concerns us is Part IV, which has the following format:

PART IV
1979 057A 09345 105066560150 810414203210007 1509616
01011681 01012254 00124732 17142454 13773458 09867899
18869440 07189253 M053313427 P048448725 M000019396
P00759127 P00825300 P07350534 003263350 245206018 9449
0000499998 M00290091 P00098722 P00512415 SPARESPARE

The nodal period of the satellite is always the second data group on line two, which I have underlined for clarity. This particular TBUS abstract is for NOAA-10. Note the nodal value of 01012254, which, not surprisingly, translates to 101.2254 minutes! This is precisely the same value we would have obtained in our earlier calculations had we rounded the value to four decimal places, as is the case with TBUS nodal period values.

Other Tracking Programs

The PREDICT program listed earlier is intended to be quick and convenient and is optimized for operation with the omnidirectional antenna system. If you need precision tracking data, there are a number of useful programs that work directly from the NASA Keplerian element sets. SAT TRAK, which has already been noted, is one example. A number of public-domain and shareware programs are available on bulletin boards. AMSAT (The Radio Amateur Satellite Corporation, an organization devoted to amateur satellite communications) has a number of fine programs. The more sophisticated programs provide real-time readout for antenna bearings on your display, while plotting the satellite position on an on-screen map. Commercial programs for many different computers are also advertised in major Amateur Radio journals such as *QST, CQ,* and *73 Magazine.* Whatever your tracking needs, you should be able to find a program for your computer system.

GEOSTATIONARY-ANTENNA BEARINGS

One of the major advantages of the geostationary satellites is that, barring the movements of any satellite to compensate for the loss of another operational satellite, once your antenna azimuth and elevation have been set, they need not be changed. In effect, its like having your TV antenna permanently aligned on your local TV station.

There are various ways to determine the proper alignment, but there is no need to be too elaborate because we won't be doing the job often, and we only need to get close enough to simply hear the satellite on the first try. As long as we can hear the satellite signal after making preliminary antenna adjustments, final fine-tuning of elevation and azimuth can be done by listening to the signal on the station receiver.

We can accomplish the job rather quickly by using our orbital-plotting board. Locate the position on the equator of the satellite you want to hear. Using the edge of a piece of paper as a straight edge, locate the straight line that intersects both your station location and the satellite subpoint. Your antenna azimuth can now be read off the bearing circle normally used for tracking.

With the paper still in place, use a pencil to mark the paper at the satellite subpoint and at your location. Now shift the paper so the subpoint is still on the equator, but the paper edge passes through the pole on the diagram. Now, read the latitude of the point on the paper edge that marked your station location. This value, in degrees (N or S is irrelevant) represents the *Great Circle arc length* (in degrees) between your station and the satellite.

With this value in hand, refer to Figure 8.7. This is a plot of the relationship between the arc distance (in degrees) and the required antenna elevation relative to the horizon. Locate your arc distance on the vertical scale, then move horizontally to the right until you intersect the graph line. If you drop straight down to the horizontal scale from this point, you can read off the required antenna elevation angle.

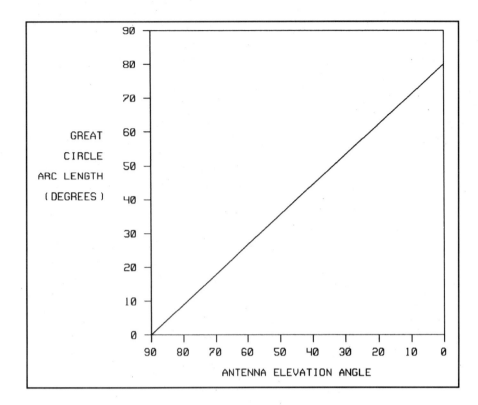

Figure 8.7—Relationship of Great Circle arc distance and geostationary antenna elevation.

Note that if the arc length exceeds 80°, the satellite will be below your RF horizon and thus out of range. Actually, any value greater than 70° arc length will put the antenna within 10° of the horizon with weaker signal levels, and raises the potential for full or partial masking of the signal by trees, buildings, hills, etc. It's a good idea to check out the bearings for satellites within range of your station before making a decision on the permanent location for the antenna. If in doubt, smaller antennas are easily mounted on a small **A** frame of lumber, so you can run actual reception checks from different locations.

Assuming the size of your antenna is in the 4- to 6-foot (1.2- to 1.8-m) range, you should be able to hear the satellite when the antenna is set to the indicated azimuth and elevation values. Remember that azimuth is given in degrees true bearing so if you use a compass, be sure to correct for your local magnetic declination. Many TVRO satellite dealers have bubble protractors to lend or rent that will help you set the antenna's elevation. A simple protractor and plumb bob, referenced to the feed-horn support mast, can also be used. If nothing is heard, double-check the satellite transmission schedule to make sure that the satellite is actually on the air. For larger antennas, you may have to tweak azimuth and elevation to get the first signal indication.

Once you have any kind of signal from the satellite, optimize the azimuth and elevation positions, and antenna polarization for optimum signal. The foregoing discussion assumes the use of the common az/el mounting system. If you are using a TVRO dish, you may have a polar mount. If this is aligned following the manufacturers instructions, you can use the position drive system to swing across the entire geostationary arc, zeroing in on each satellite by adjusting for maximum signal levels. If you take this approach, set your feed horn for vertical polarization, with the dish facing south, and no further polarization adjustments should be required. Once the satellites within range have been located, you can program their positions into the TVRO dish-position controller.

If you have a TVRO installation, you can often mount the S-band feed horn (facing the dish center) off to one side or the other of the TVRO feed. Do not mount the horn above or below the feed as this would require that the dish elevation be changed—a most undesirable option once the polar mount has been set up!

Chapter 9

Station Operations

SAFETY

One advantage of modern solid-state equipment is that you don't have to deal with potentially lethal high-voltage supplies. A single UL-approved 12-V power supply in the 4- to 5-A range will handle virtually all your equipment needs. Even with allowances for the computer and lighting, the current demands of your installation should be quite modest. All the ac power needs for your equipment can probably be handled by a single power strip. This unit should be equipped with its own circuit breaker. It is also handy if the strip incorporates built-in surge suppression, to protect your equipment from transients and surges on the ac mains. Alternatively, a surge suppressor can be installed where the cable from the power strip is plugged into the mains socket.

All equipment should incorporate either polarized or grounded, 3-wire power cords. These should not be modified in any way. If your station location does not have 3-wire ac mains sockets, it is only common sense to have these installed by a licensed electrician. Power cables should be neatly routed so they cannot be stepped on or tripped over. If children have access to the station, particularly very young children, unused ac receptacles should be equipped with inexpensive plastic shields.

Good grounding is an asset to any installation, both for safety and to reduce the potential for RF interference from your computer or other sources. The chassis ground of all equipment should be tied together with heavy-gauge wire. Ideally, this common ground bus should then be connected to a solid earth ground. A cold water pipe or an external ground rod (driven at least eight feet into the ground) is adequate. If your house employs PVC pipe, verify that your cold water pipe actually has dc continuity with the water main coming into your home.

Having a small CO_2 or Halon fire extinguisher handy is also a good idea. If you will be building equipment on site, be sure that the soldering iron is placed so that it cannot come into contact with combustible materials or visitors. A master power switch that will turn off all bench equipment, including the soldering iron, will minimize the chances of the iron being left on for extended periods.

STATION CONFIGURATION

Not many years ago, a weather satellite installation required lots of space. Bulky fax machines, surplus receivers, tube-type power supplies, and even a darkroom for processing pictures were all necessities if you wanted to view the earth from space. Today, with the shift to computers for display and the availability of compact, solid-state receivers, a satellite station can fit almost anywhere. While working on this edition, I finished the development of the WSHFAX program on a PS/1 computer in the kitchen that functions as the Taggart Family Computer Center. I needed access to live satellite passes at several stages in the process, so I added a Vanguard WEPIX 2000 receiver and the prototype WSH Satellite Interface to the corner alcove. The result was a complete weather satellite station, tucked in the corner of the kitchen. The family complained about the piles of papers, magazines, books, and reference manuals that cluttered up the corner, but they were completely unaware that I had slipped in another weather satellite station when no one was looking!

The point is, the station can be as simple or elaborate as your situation dictates. It does pay to have some space to spread out when caught up in the throes of a new experiment. In my case, "spread out" is a euphemism for chaos, which is why you will never see a picture of my basement workshop! At the other end of the scale, there are neatly integrated stations where everything is neatly packaged in a console, with all the appearance of a state-of-the-art aerospace installation.

While I would never encourage anyone to emulate my workshop, there are problems with packaging things too neatly. It can make it difficult to pull things apart for testing, if a problem develops, or to do some experimenting with new modes or equipment. If you have to pack everything into a relatively tight space, giving some thought to switching options can do a great deal to improve the flexibility of your station.

Proper switching for audio, computer parallel cables, and RF transmission lines can make it very easy to alter equipment configurations without having to pull everything apart to get to inconvenient input and output connectors. If you plan a switching system, make sure that you have easy access to the receiver output and interface input, the computer's parallel port, and the receiver RF input as well as the signals from the various RF transmission lines. While an actual set of switchboxes or panels is the most elegant solution, a combination of switching and an RF patch panel can provide the ultimate in flexibility.

TAPE RECORDING

The bandwidth and audio-frequency characteristics of the various satellite signal formats give us a major advantage—the ability to record satellite signals on audio tape equipment. In the past, this was accomplished using ordinary stereo tape decks, but that approach involves a number of potential problems.

The first problem is speed regulation. While the speed of any stereo tape system is adequate for high-quality audio reproduction, the speed of recording and playback *does* vary slightly. The resulting timing errors are sufficient to destroy the quality of the image. This problem can be solved by recording a sample of a crystal-locked audio tone (such as the interface CLK signal) on one channel of a stereo tape deck, while recording the satellite signal on the alternate channel. On playback, the recorded reference tone is used to lock-up a phase-locked loop (PLL) tone generator that tracks the frequency variations in the reference tone caused by recorder speed variations during recording and playback. If the variable PLL tone is used as the display frequency reference, the display will stay in synchronization, despite inevitable speed variations in the tape system. This fix adds to the complexity of the system and also presents some compatibility problems. Different satellite systems will often employ different reference tone frequencies. In practice, this means that images recorded using one display system may not be compatible with another display.

A more serious problem is the quality of the recorded image data. Almost anything can degrade the recorded signal, introducing all sorts of problems with the final display. Most of the problems involve changes in the peak signal amplitude. The changes are caused by variations in the tape oxide layer, the quality of the tape, and other problems associated with the recorder itself. The result is typically an image with a multitude of dark streaks resulting from momentary amplitude variations.

What is needed is a recording medium that does not have speed instability. It turns out that there are two viable approaches that can yield excellent results. The optimum solution is a digital audio tape (DAT) system.

These decks use digital-audio technology and record information on audio cassettes that are quite similar to those used in 8-mm video equipment. A DAT recorder can make a perfect reproduction of the original signal and do it without appreciable speed variation in record or playback. A DAT deck is the ideal solution to tape recording of satellite signals, but there are two practical difficulties with this approach.

First, the systems are still quite expensive, with low-end machines in the $600 to $1000 price range. Prices have come down and can be expected to drop still further as market volume increases but that brings up another issue. *Will* market volume increase? Introduction of consumer DAT systems was delayed by all sorts of legal gymnastics by the recording industry. Systems are available now, but so are rumors of a new generation of record/playback audio CD technology. No one is eager to shell out a lot of money for equipment that may become obsolete before it really has a chance to catch on! There is no doubt that affordable consumer digital audio systems will arrive in the near future, but the format is very much up in the air. Had DAT come on-line a few years earlier, its position might well be assured. The issue is now very much in doubt. Make no mistake, a DAT machine will serve you very well—if cost and long-term commercial viability are not your major concerns.

Fortunately, there is another solution out there in the form of the ubiquitous video tape deck! When a VCR plays back a recorded video tape, an internal servo system locks to the recorded TV synchronization data. This allows the VCR to maintain the accuracy of the playback speed while minimizing any variation. If the system didn't so this, it would be virtually impossible to reproduce the TV video signal. Since the same circuits maintain speed accuracy during recording, we have a tape system that maintains more than sufficient accuracy for satellite image display. If the satellite signal is recorded on the audio channel while a TV image is recorded on the video channel, the satellite signal will maintain perfect synchronization on playback.

The way that audio track signals are encoded for recording on a VCR varies greatly. Some decks use a method essentially identical to ordinary audio recording, while others use elaborate frequency or pulse-code modulation techniques. The resulting audio quality can range from identical to a good-quality audio machine, to dramatically superior. *Any* VCR can do a very good job of recording satellite signals, but some formats are really excellent. For example, 8-mm camcorder systems use a rotary-head FM audio modulation system that virtually eliminates problems with signal drop-out. I use a Sony 8-mm video tape deck for recording satellite signals and get superb results.

The key to successful recording on a VCR is to make provisions for recording a TV image on the video track while recording the satellite signal on the audio track. This can be accomplished in a number of ways. In the case of a typical VHS VCR with a built-in tuner, connect a simple antenna to the VHF input (or leave it connected to cable) and tune in a strong local signal. Connect the output of the satellite receiver to the **LINE AUDIO** input of the tape deck and the interface to the **LINE AUDIO** output. Now when you set the system to record, the TV video signal will maintain the speed accuracy of the system while the audio track is recorded. Alternatively, the video output of a TV camera or camcorder can be routed to the **LINE VIDEO** input to accomplish the same function.

With a little shopping you can find a new VCR at a price that is competitive or even lower than a high-quality audio system. This allows you to dedicate a VCR to your satellite station. If you need to rationalize the purchase, tell the family that it is an emergency backup against the day that the family VCR gives up the ghost! Of course, appropriating the family VCR is a valid alternative if you only want to make occasional tapes.

UNATTENDED RECORDING

It is a sad fact of life that some of the most interesting satellite passes occur during the day when, for five days out of every seven, you are hard at work paying for all this nonsense! One solution is to make recordings of passes while you aren't at home. One of the advantages of using a VCR for recording is that the recorder has the clock and the hardware to make recordings according to both the day and the time. With time data available from your tracking program, you can program the system to make the recordings.

In the case of the WSHTRAK and PREDICT programs, if you set the system to come on at the indicated AOS time and switch off at LOS, you will have bagged the pass with enough leeway at either end to assure that nothing is missed. It helps to set the VCR clock to accurate GMT/UTC time so you will not make mistakes due to time conversions.

The widespread use of computers for image display has greatly reduced the need for unattended recording. Many systems will log multiple passes, storing the image data to disk for later playback. The WSHFAX program will do this, but it does have limitations. For example, you can select visible or IR data for a TIROS/NOAA pass, but not both. Also, if noise should disrupt the automatic phasing routine at the start of the display, it is possible that the desired data channel may be improperly phased. Although computers provide an excellent alternative to recording, a tape provides access to data from a pass as well as the opportunity to display the image as many times as you wish.

If you are experimentally inclined, there is an excellent opportunity to combine the capabilities of the computer with your receiver and VCR. For example, a program like WSHTRAK can evaluate the passes for all spacecraft on file for any given day. It can evaluate the pass in terms of the signal quality and it even knows the operating frequency of each spacecraft. The computer running the program has a parallel printer port with eight printer output data lines, four usable control lines, and five input lines. A receiver, like the WEPIX 2000 controls frequency using an EPROM with eight data-output lines. It is quite possible to use the printer data lines to control the receiver frequency!

Your typical VCR is also a digital device, with all switch functions—including record and stop—implemented as individual data bits. There is no reason why the control output bits couldn't be used to start and stop the VCR under computer control. You could implement a hard-wired connection, use the computer to control the IR remote, or even use the computer to key an IR LED to control the VCR. Combine these ideas any way you like to achieve a tremendous range of flexibility. A typical video tape, operated at standard speed, will record two hours of data. Assuming 10 minutes of prime data per pass, such a system could easily log 12 passes on a single tape. That is a lot more data than you could expect to log directly to the hard-drive of most computers, yet it requires almost no additional hardware. What's more, this application is *not* speed intensive and thus could be implemented via simple add-ons to a BASIC program like PREDICT. If you are looking for a low-cost experimental activity, computer control of the receiver and/or VCR is an excellent project.

HARD COPY

One of the advantages of scan conversion is that once you've made your basic hardware investment, you can watch any number of pictures without additional expense. A CRT display monitor requires that each image be photographed, a negative processed, and a print made to see the final results. The costs are not insignificant:

Film/frame:	$0.15
5 × 7 paper:	$0.24
Chemicals:	$0.10
TOTAL/print	$0.49

This does not, of course, include the value of your time (or someone else's) in doing the processing. You can, of course, use Polaroid instant film, but the price you pay will be about $0.75 per image with a maximum print size of 4 × 5 inches. Time and economics definitely limit the number of pictures you can look at.

Fax recorders will print the image directly, either on photographic paper or on specialized electrostatic or electrolytic papers. Photographic systems will average about the same cost per image as the CRT system. (Even though you skip the film-processing step, you will be using larger paper sizes.) Electrostatic or electrolytic printouts may cost as little as $0.10 per image but the image quality will not be as high and some processes are not completely permanent.

Although the use of a computer can eliminate the cost of viewing pictures, saving pictures does cost something in terms of the tapes or disks used to save the image. For example, I can store 10 of the 512×480 images you see throughout this edition on a single 1.44 megabyte 3.5-inch floppy disk. Since the cost per disk is about $2, I have put out about $0.20 per image for the many hundreds of images I have stored in disk files. Other media cost less, but the per-image storage cost does not vary greatly since the number of images per disk is also reduced.

Despite the advantages of disk storage and recall, there will always be the need to produce some sort of hard copy. The production of this book is one case in point. Our two viable paths to get from a digitally stored picture to a permanent print involve either photography, or the use of a computer printer.

Photography

Getting good photographs from either your TV monitor or high-resolution computer display involves a few tricks, but the process isn't difficult. First, you need a good camera. A single-lens reflex (SLR) camera for 35-mm film is the best choice. Most standard 50-mm lenses will allow the camera to focus on a 12-inch display. Many macro lenses will allow you to fill the frame with a smaller monitor screen. We want to fill as much of the image area as possible so we can produce crisp enlargements without worrying about the grain of the film.

One key to taking good photographs is to keep in mind that it takes a finite length of time for the monitor to *paint* the image on the screen—about $1/60$th of a second. Any shutter speed faster than this value will produce a partial scan! The situation is even more complex due to the nature of the focal-plane shutter system used in typical SLR cameras. The shutter acts as if it were a window shade, snapping out of the way to expose the film and then snapping back into place to interrupt the light from the lens. As it moves during the course of the image scan, the interaction of the raster scanning and shutter position will leave a strong diagonal pattern on the screen. There is no way to avoid this, so we must make the total exposure rather long compared to a single scan of the screen. With long exposures, the effect becomes unnoticeable. I would suggest exposures of $1/2$ to 1 second. An exposure of this length mandates the use of a camera mount (such as a tripod) to hold the camera steady during the exposure period. Also, room lighting should be very subdued so ambient light does not wash out the image and reduce contrast. The room doesn't have to be completely dark, but it should be dark enough that the display screen looks quite bright to the eye.

Since the exposure length is set by the problems inherent in focal-plane shutter systems, the only other variable is the lens opening, or *f-stop*. Kodak Plus-X panchromatic film is an excellent film that offers good speed (ASA 125/DIN 22) and fine grain. Using this film with an exposure of $1/2$ second, the required f-stop will usually be in the range of f16 to f8. Good image contrast is achieved using f11 with both my TV monitor and the VGA display from my IBM PS/2. You can get an approximation of the required f-stop by taking an exposure reading from a full-screen grayscale display. You can also take a series of test shots of several normal pictures using a range of f-stops, recording the lens opening for each shot. The negatives or prints from this test roll can then be used to pick the optimum f-stop for your display. The monitor brightness and contrast settings should also be recorded so that you can achieve the same results each time.

Most of the photographs that illustrate this edition were taken with a vintage Pentax camera using Plus-X film. Films were processed in Kodak Microdol X developer (seven minutes in a standard roll film-developing tank) and fixed with Kodak Rapid Fixer. Prints were made on Kodabrome RC paper (normal contrast) processed in Kodak Dektol developer and Kodak Rapid Fixer. All of my darkroom facilities are very basic (my total investment is about $200), but I prefer to process my own photographs since it gives me total control over the final product. You can use commercial photo processing, but expect the process to take about a week since most photofinishers do comparatively little black and white work these days.

If speed is important, the images can be photographed using color print film, with the option for commercial processing in as little as one hour. The photographs that illustrate this chapter and Chapter 10 were obtained this way. While not equal to the quality that can be obtained in your own darkroom, the results are often acceptable.

Computer Printouts

There are innumerable interfaces and software systems out there that advertise fax capabilities. Most of these are designed to provide hard copy using a dot-matrix printer. Such printers can do an excellent job on weather charts transmitted on HF, or even the charts transmitted on WEFAX. The reason they can function so well is that a chart or other printed material (TBUS messages, schedules, and the like) represents binary data. A given pixel is either black or white. In looking at the problems of printing an image, the

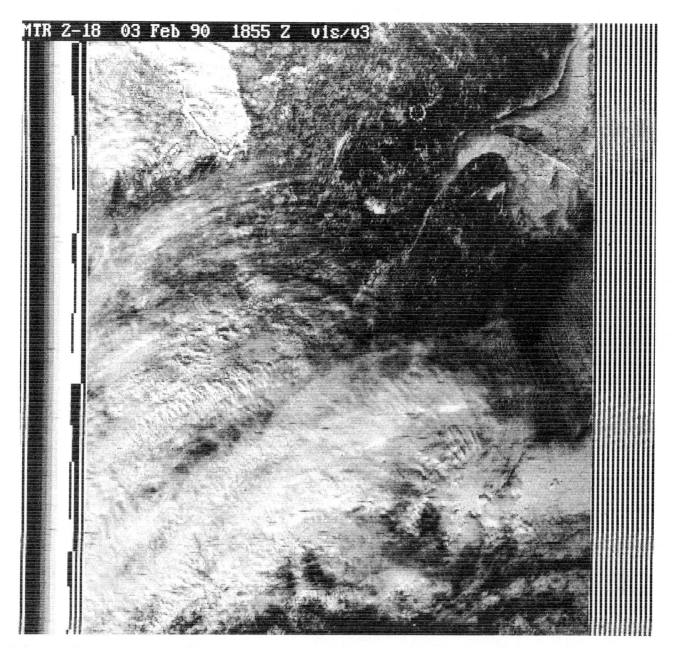

Figure 9.1—Computer printing technology has advanced to the point where your printer can generate reasonably good copies of images stored to disk. This is a METEOR pass over eastern Canada displayed using the WEFAX program for the interface unit in Chapter 5. The image was printed on a Hewlett-Packard DeskJet printer using the <J>et-print option from the display menu. This routine supports image printing on any HP LaserJet-compatible printer, including the relatively inexpensive DeskJet.

printer control program can either print a dot on the paper (for black) or not print (resulting in white). Virtually any dot-matrix printer with high-resolution graphics capability can print such material with good fidelity.

In order to get reasonable fidelity in a typical satellite image, we need to do far more than print simple black and white. We must be able to print varying grayscale shades—at least 16—and do so without degrading the spatial resolution of the image. For virtu-

ally all dot-matrix printers, creation of the *dither patterns* necessary to reproduce grayscales takes up space on the paper. The result is very much like looking at a newspaper photograph using a magnifying glass. Viewed up close, such an image is textured in the extreme and looks good only when viewed from a distance. The images will be interesting—even fascinating—if you've never seen the pictures properly displayed. The typical grayscale dot-matrix printouts are a tribute to the programmer's skill, but they are nothing more than a crude representation of the actual image.

Modern laser and ink-jet printers are another story. At relatively low resolutions of 75 or 150 dots per inch, they are no better than the typical dot-matrix printer. But at 300 dots per inch, using a decent driver program, they produce results (Figure 9.1) that rival many fax machines.

Doing this job with a laser printer can be expensive. First comes the cost of the printer, the expense of toner cartridges and other supplies, and, worst of all, the cost of anywhere from 1 to 3 megabytes of RAM necessary to do a full page of 300 dot-per-inch graphics! Prices have been coming down steadily, but you can still suffer sticker shock.

A far less costly alternative is the Hewlett Packard DeskJet printer. This printer does not require large amounts of RAM and uses simple liquid-ink cartridges. With careful shopping, you can usually obtain one for about $500. While this is expensive compared to the $200 or so one might expect to pay for a 9-pin dot-matrix printer, it is not out of line with the price of better 24-pin printers. Two other factors also require consideration. First, the DeskJet is virtually silent. If you compare it with the incredibly annoying racket of a dot-matrix printer working in a graphics mode, there is no contest. The second factor is reliability. The better dot-matrix printers have an MTBF (Mean Time Before Failure) rating of 4000-5000 hours. The DeskJet's MTBF rating is 20,000 hours! The DeskJet is completely LaserJet compatible, so any LaserJet graphics program will drive the DeskJet. In fact, I actually prefer the output of the DeskJet in these applications.

Most graphics programs support printing routines. Many even have screen-capture functions that will let you "grab" one of the file images without having to use the WSHFAX <E>xport function (discussed later). WSHFAX does support Laser and DeskJet printers. Many of the printing routines in the graphics programs are faster than the WSHFAX routine, but I have yet to find one that's better! Remember, you don't have to have a laser printer—you just have to scrounge time at odd intervals on a computer that *is* connected to such a printer!

SATELLITE BULLETIN BOARDS

Once you begin to reach operational status with your station, you need to keep up with a veritable flood of information. Orbital elements, new spacecraft, the status of existing systems, hardware and software evaluations—the list seems endless. The most effective way to accomplish this is to add a modem to your computer and take advantage of the wealth of information available from bulletin board systems (BBSs).

The premier BBS for weather-satellite enthusiasts is the one operated by the Dallas Remote Imaging Group (DRIG), spearheaded by the SysOp, Jeff Wallach. DRIG is dedicated to providing the most up-to-date information on weather satellites, remote imaging, Amateur Radio satellites, and the space program in general. It would take a full chapter just to outline the tremendous range of resources provided by this bulletin board. DRIG provides all the data you need to keep track of both the US and Russian polar orbiters (not to mention just about anything else in orbit) and they have a FILES section that boggles the mind. Public domain and shareware programs for satellite tracking, screen capture, high-resolution image display on all the major computer systems, and an unparalleled library of APT, HRPT, and VAS images. Here you'll find new products announced and reviewed. The electronic mail section provides you the opportunity to ask questions that will be answered by some of the sharpest folks in the business.

If you've gone to the trouble of putting together a satellite station and then fail to get a modem and connect to the DRIG system, you are missing out on the most current source of information available to you. If there is a better source of satellite information on planet earth, let us all know about it! Jeff Wallach wrote Chapter 9 on HRPT and VAS high-resolution satellite systems and has included additional information on the BBS. There is also a DRIG ad at the back of this edition.

A much more modest effort is my own WSH BBS system. This is a free system devoted primarily to supporting WSH projects and helping readers with their problems. The BBS is informal and there is no charge for using it. You can reach the WSH BBS at 517-676-0368 using 300-2400 baud, 24 hours a day, 365 days a year.

A number of vendors, many of whom are advertisers in this edition, maintain their own BBS systems. These can be very valuable sources of information

IN CASE OF DIFFICULTY

Taken as a whole, the current system of weather satellites is terribly complex. There will inevitably be times when you begin to notice problems. Your first step should be to thoroughly check the various components of your ground station, using other satellites if necessary. If you are having similar problems with a number of spacecraft, the likely source is your own installation. Even if all the electronic systems are functioning, there is always the possibility of water in connectors (or transmission lines) or other subtle problems that need checking.

If the problem seems to be with a single spacecraft, some other factor may be involved. Use the electronic bulletin boards to see if others have noted the problem or offered solutions. Only when the problem can be definitely assigned to the spacecraft or the ground support system should you proceed to check with the folks at NOAA. The amateur community has acquired a high degree of credibility over the years and there is no need to compromise it by complaining about problems that cannot be verified from several sources. We provide a valuable monitoring function that can be a real service to the technical agencies that operate the spacecraft. Even so, it is imperative that our reports be absolutely reliable. While you are still relatively new at this game, double check your observations with others to avoid embarrassment. There is a huge reservoir of talent out there on the bulletin-board systems. With all of us willing to pitch in, your problems can probably be localized and corrected in short order.

THE WSHFAX PROGRAM

In many years of writing software, I have discovered that writing the documentation is a lot of wasted time. Ninety percent of the questions I receive are covered in the documentation—which nobody seems to read. Since most of you will read this book several times while working to get your stations on line, there is a very real possibility that you might actually become familiar with the program before you get to use it! That will save both of us a lot of time and will let you have your system up and running as soon as it's finished. Outlining the various program features also provides would-be programmers some guidelines. Don't hesitate to steal some good ideas if you are thinking about programming for the interface!

By the time you finish this section, it should be obvious that WSHFAX is *not* a watered-down "demo" program intended to entice you into spending more money on a really functional software package! I noted in Chapter 5 that I designed an imaging card for the PC that was an award-winner in the Zenith Data Systems *Masters of Innovation IV* competition. The project was a winner because it combined an innovative hardware component—the imaging card—with a very nice software package. As noted in Chapter 5, I did not include the card in this edition, but that doesn't represent a compromise. As good as the card was, the WSH Interface in Chapter 5 is better functionally, in addition to being far more flexible in application. The WSHFAX program I wrote to support the interface is better as well. It combines all the best features of the imaging card software, but is easier and more flexible to use. Don't let the modest price fool you—it will do just about anything that most of you will want to do (plus lots of things you probably never thought of!). This is, after all, a hobby and not a national security project. There are things the program won't do, like overlaying geographic grids from government data bases, measuring temperatures from IR data, or doing elaborate false-color displays. Actually, you can do false color since you can export image files in the universal importing to "paint" programs for colorizing, labeling, or anything else you can think of. My philosophy is simple—there is no point in making 100% of you pay for exotic features that only 5% will ever use! If you have very specialized needs, the many advertisers in this edition have products in all price ranges that will probably do just what you want.

WSHFAX covers all the basics of an image display program and includes enough bells and whistles to make it an extremely powerful application. It can serve as a useful guide to the kind of functions you might wish to incorporate in your own software. Once you have crafted a basic functional package, you can then experiment with lots of useful enhancements, as well as customizing your package to conform to your ideas about how the system should function.

HARDWARE REQUIREMENTS

For openers, the software will run on any PC/XT/AT/PS1/PS2 compatible with at least 640k of RAM, one floppy drive, a parallel port, and a VGA display. With that said, I have to be up-front about the kind of system you will need if you really want to enjoy the program. Other programs you might use are likely to be even more demanding in terms of system requirements.

The CPU

Your basic PC or XT, in an unmodified form, is probably running an 8088 or 8086 microprocessor (CPU). The basic problem with these processors is their speed limitations. Four- and 8-MHz systems will probably be too slow to keep up with the satellite data stream in terms of real-time imaging. The bottleneck is writing data to the VGA screen. The BIOS routines in Chapter 6 take too long with these processors. The clock speeds and the computer will fall behind, resulting in a chaotic screen as opposed to a well-organized image. The screen video routines in WSHFAX write directly to the VGA registers (a real pain for the programmer, but much faster than BIOS), so a 10-MHz "turbo" system *may* be able to keep up. It will depend on the computer and operating system.

The 80286 processor, first used in the AT (6-MHz clock), is a major improvement. A 10-MHz 286 system (bottom-line by today's standards) will do just fine with WSHFAX. Given the fact that 286 machines are now considered stone-age, you can certainly score a heck of a deal on a used system.

If you buy a system today, it will almost certainly be equipped with a 386 or 486 processor (SX or DX). There is no chance that the hardware will limit your ability to use WSHFAX or even more advanced applications built around the interface.

System RAM

The software absolutely demands 640k of available RAM. If there are any fossils out there that are still limited to 256 or 512k, you will have to bring it up to the 640k DOS limit. Most of today's entry-level systems start with 1, 2, or even 4 meg, so memory shouldn't be a problem in most installations. The additional memory doesn't matter in terms of WSHFAX, but is immensely useful with other applications.

Disk Drives

The WSHFAX program doesn't care if it is operated from a floppy-only system, but the capacity of the drive has a major impact on your ability to store images. The program supports two image-storage formats: the standard FILE format (128k per image) and the BINARY format (256k per image). The following represents the maximum image storage capacity of the common floppy drive formats:

Drive	FILE	BINARY
360k	2	1
720k	5	2
1.2 Meg	9	4
1.44 Meg	11	5

Given the program's requirement to use the disk drive for temporary data storage, plus severe limitations on file storage capacity, a 360k drive is not practical. 720k is a realistic minimum, but won't be fully functional without a 1.44 meg drive. The other problem with smaller capacity drives is that they tend to be very slow. This won't impede program functions, but it's a real time-waster.

A hard-drive user is a happy user, both in terms of speed and capacity. Bigger is better in the size department.

VGA Display

In order to be as universal as possible, the WSHFAX program uses the 640 × 480 mode for image display. Super VGA cards and monitors are now available that will do a 1024 × 768 display and these are becoming common. This level of resolution is pushing the theoretical resolving limit of most satellite modes, so there is no point in getting a system with more resolution. WSHFAX won't take advantage of the higher resolution, if you have the capability, but other applications do and future releases of WSHFAX will probably do so.

WSHFAX sets up the display adapter for a grayscale display. Therefore, a monochrome VGA monitor will work just fine if funds are limited. A color monitor will display WSHFAX images in grayscale, but it also gives you the option of playing with false color (see Chapter 10).

Any of today's entry-level systems are usable without imposing limits on what you can accomplish. Anything less effective than a 286 (10 MHz) with a modest-sized hard drive will probably require upgrading if you hope to be completely happy with the system!

GETTING WSHFAX UP AND RUNNING

The following instructions assume that you are running DOS. The equivalent functions can be performed from an operating system like Windows or OS/2, but you must exercise your mouse!

Program and File Installation

In keeping with a program that is supposed to be easy to use, installation of WSHFAX is extremely simple. Your program is contained on a distribution disk (probably a 3.5-inch, 1.44-Mb floppy) containing WSHFAX.EXE, at least two sample FILE images and two BINARY images (more on that later). The following assumes that you will be working from drive A and installing on hard drive C. If this isn't the case, make the appropriate changes in the DOS commands. In all cases, <ENTER> means to simply hit the ENTER or RETURN key. First, make sure you are in your root directory by typing:

cd\ <ENTER>

Now create a subdirectory, WSH, as follows:

md \wsh <ENTER>

Now move to the new directory by typing:

cd \wsh <ENTER>

Now insert your distribution disk into drive A and copy all files as follows:

copy a: *.* <ENTER>

That completes WSHFAX installation. You may also find it convenient to copy your tracking program (PREDICT, WSHTRAK, or whatever) to the WSH directory so you can run them without changing directories. Be sure to copy any relevant configuration and data files as well.

Running WSHFAX

The simplest approach to booting WSHFAX is to simply change to the WSH directory:

cd \wsh <ENTER>

and then boot the .EXE file:

wshfax <ENTER>

You can streamline the process even further by creating a simple batch file (WSH.BAT) using your word processor in the nondocument or ASCII mode:

cd \wsh wshfax cd \

If you place a copy of this short file in the root directory, and in the directories for your commonly used applications, you can boot the program from almost anywhere in your system by typing:

wsh <ENTER>

As scripted, the listing shown above returns you to the root directory when you exit the WSHFAX program. To return to another directory, simply include the path data in the last line. To exit to the WSH directory, simply omit the last line.

A Note on Conventions

Most program functions are menu-driven and implemented with single keystrokes of one of the alphanumeric keys. For example:

<L>oad File

would be implemented by keying L. Although the options are shown in uppercase, the choices are *not* case sensitive. As with most applications, it is easier to ignore the <SHIFT> or <CAPS LOCK> functions and use lower-case input. There is no need to use the <ENTER> key—simply hit the indicated key and the function will be implemented.

WSHFAX does not support mice. This not a drawback because you will be implementing functions faster with single keystrokes than you ever could by pointing and clicking. Also, you don't need a pad or clear table space.

The only occasion where you need multiple keystrokes and the <ENTER> key is where you supply filenames for loading or saving images. Extensions are never used when handling WSHFAX files from within the program (they are appended automatically, depending on file type), so all you have to assure is that the name you use follows the DOS file conventions—eight characters maximum with no spaces and some limitations on available characters.

Parallel Port Check

The first thing that will happen when you boot the program is a systematic check of the three possible port addresses to find the interface. This means that you can move your interface to different parallel ports or between computers, without worrying about configuration files. Once the interface is located, the program will display the Main Menu.

If the program cannot find the interface, an error message will be posted, along with suggestions as to where the problem may be. Assuming your interface passed its check-out, the most likely problem is that you forgot to turn it on, or that it was disconnected from the port for some reason. If the problem is obvious, fix it and then initiate a retest by keying <T>.

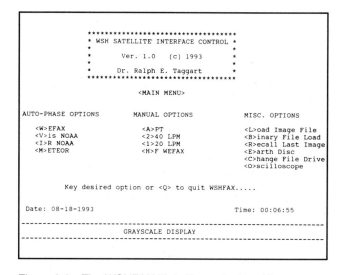

Figure 9.2—The WSHFAX Main Menu display. All commands are executed with single keystrokes. The menu also provides real-time display of time and date.

If you wish to continue (despite the fact that the interface has not been found), simply key <C>. The program will not let you acquire images (since that requires access to the CLK signal), but you can use all other program functions. That lets you install the program on other computers and use them for manipulating image files, demonstrations, etc.

If you have a problem and don't wish to continue after the failure of the port test, simply key <Q> from the error screen to quit WSHFAX.

THE MAIN MENU

The Main Menu (Figure 9.2) is the crossroads of the entire program. Exiting from virtually all functions (universally implemented with the <Q> key) will return you to this display. To get out of WSHFAX from the Main Menu, simply key <Q> and you will be dropped back into your operating system.

Note that a 16-step grayscale is posted along the bottom of the screen. If the monitor **BRIGHTNESS** and **CONTRAST** are properly adjusted, all 16 steps will be visible, without blooming or trace distortion. Also note that the system posts both the date and time, which are updated in real-time. This is a real aid in tracking a spacecraft. If the computer clock is set to the GMT/UTC date and time from DOS, you don't have to worry about making mental time corrections and you will eliminate the need for a station clock.

If the system does not receive keyboard input, the screen will automatically blank as a screen-save function after about two minutes. Hitting any key will restore the display.

Note that the Main Menu functions are organized into three major groups: AUTO-PHASE IMAGING, MANUAL IMAGING, and MISC. OPTIONS. Although you are understandably eager to turn those noises from the receiver into real pictures, let's start by exploring the MISC. OPTIONS functions so you will know what to do when you do get your first images!

<L>oad File Image

"File" images are the standard image file format for WSHFAX. They combine high quality with modest file size, making them ideal for long-term storage. Key <L> and the screen will clear and you will be presented with a directory listing of all file images in the WSH directory. At least two sample file images were provided with your distribution disk. Note that file images all have a .480 extension. To load one of the images, type the filename (*no* extension) and hit <ENTER>. The image will be loaded in a few seconds. If you wish to exit from the directory without loading an image, type <Q> and <ENTER> and you will return to the Main Menu. If you proceed to load one of the two images (I suggest that you do so now), you will be routed to the FILE DISPLAY screen. I would suggest that you jump ahead to the section titled WSHFAX FILE DISPLAY so we can explore the options you now have available.

<B>inary File Load

When you acquire a satellite image, the image data in the RAM buffer are coded to eight bits, representing 256 possible grayscale values. Translating 8-bit RAM data to 4-bit display data provides a huge range of options for processing the picture. The Binary File format provides a quick way to save the contents of the RAM buffer so you can experiment with the picture later. When you key <B> from the Main Menu, you will be presented with a listing of all the available Binary Files. Two have been provided with your distribution disk and should appear on the file list as VIS_RAW.A and IR_RAW.A. To load one of these images, type the filename (*no* .A extension) and hit <ENTER>. In moments, the file will be displayed, along with a range of options, under the heading WSHFAX IMAGE DISPLAY down the left edge of the screen. The IMAGE DISPLAY screen is the same screen you will be using when you display your own satellite images. In fact, having loaded a Binary File, as you just did, you have precisely the same options as you would when you display an image from your own receiver. I suggest that you use this opportunity to jump ahead to the discussion headed WSHFAX IMAGE DISPLAY, to explore the options.

<R>ecall Last Image

This option lets you recall a binary image after a return to the Main Menu. As long as you have not

Figure 9.3—The earth disc, as scanned by the GOES visible-light sensors. This image was constructed automatically by the computer, based on four file images of the primary WEFAX quadrants, using the <E>arth Disc routine from the WSHFAX Main Menu. The days of patiently taping together photographs or fax printouts are definitely over.

started to load a new image or exited the program in the interim, the last Binary File or real-time image you displayed will reappear. Keying <R> is all that is required to recall the image. You can try it now if you like. This option does *not* work with File Images. If you key <R> without loading a Binary File, or having acquired a new image, the display will simply show the default contents of the RAM buffer. To look at a File Image again, you will have to reload it using the <L>oad File Image option.

<E>arth Disc

This is a neat option. It allows you to reconstruct an image of the entire earth if you have a File Image of each of the primary image quadrants. If you key <E>, you will be asked for the filenames for the NW, SW, NE, and SE quadrants, respectively. The program will then display each in turn, sampling the displayed image and using the data to reconstruct the entire earth disc. Figure 9.3 shows such an image. While less detailed than the individual quadrants, the images are striking and make superb demonstrations.

This option requires that you have the four quadrants stored as File Images. You *can* use it with any four designated files that do not make up the earth disc, but the result will simply be a montage of the four files you designated.

<C>hange File Drive

When you go to load a File Image, the files that are displayed on the file list are those in the WSH directory. By keying <C>, you can change this default to display filenames in any directory or on any drive in your system. For example,

a: **<ENTER>**

will change the default to the A drive. Typing

\archive\ **<ENTER>**

would change the default to a subdirectory called ARCHIVE. To change back to the default (directory WSH), simply key **<ENTER>** at the prompt.

<O>scilloscope

You have already used this option in setting up your interface. It displays the waveform of the image signal, much as you would see if you connected an oscilloscope to the output of the FM or AM demodulators. It has two very important functions as part of your day-to-day operations.

In the case of satellite reception, you would call up <O>scilloscope prior to image display to assure that you have an optimum setting for the SAT CONTRAST control. Key <O> and observe the display as the satellite signal rises out of the noise at the start of a pass. The effects of the noise on the signal will be clearly evident. As the noise drops out, in the case of a TIROS/NOAA pass, you will see two prominent peaks. These will occur, somewhere along the display line, representing the white pulses in the visible and IR sync pulses. *Carefully* adjust the SAT CONTRAST so these peaks *just* reach the white threshold. Figure 9.4 shows the display of a NOAA signal with SAT CONTRAST properly adjusted. At this point you can key <Q> to exit and proceed to display the image. In the case of a METEOR image, use the white peaks of the obvious sync pulse to make this adjustment. WEFAX images typically have white areas framing the image or you can use the phasing interval signal to make the adjustment.

In the case of HF WEFAX display, the <O>scilloscope function is used as a tuning indicator. Most of the time, HF FAX stations like NAM are transmitting maps or charts, consisting of black lines on a white background. If you key <O> and tune in the fax station, with the AM/FM selector switch in the FM position, the waveform of the detected signal will be displayed. The station is properly tuned when the white and black peaks *just* reach their respective limits on the display. This is the most accurate tuning

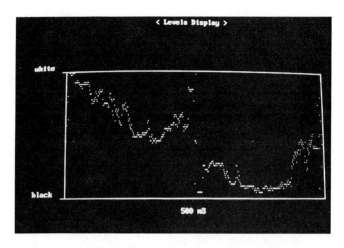

Figure 9.4—The waveform of a NOAA 12 signal, displayed using the <O>scilloscope function from the WSHFAX Main Menu. The primary function of this routine is to permit precision adjustment of the SAT CONTRAST control for satellite data (or tuning HF WEFAX signals). Two peaks, representing the visible-light sync pulse, the IR sync pulse and white pre-earth scan are visible. SAT CONTRAST is properly adjusted when these peaks *just* reach the white (upper) threshold, as shown here.

aid possible for these stations and will assure precision tuning every time.

WSH FILE DISPLAY

Assuming you've loaded one of the file images, the image will be framed on the right side of the screen. A series of options listed under the heading WSH FILE DISPLAY will appear down the left side of the screen. The various file image options are as follows:

<S>ave Image

It may seem odd to have a <S>ave function for a menu that deals with images that have already been saved, but there are plenty of reasons for the option. For example, the FILE DISPLAY menu has a labeling function (to be discussed shortly). You may have saved the image without a label, or you might want to change the label. Once you have done so, the image can be saved back to the old filename (replacing the original), or saved under a new name. You can also copy the file to another directory or drive by prefacing the filename with the appropriate DOS path conventions. Although .480 file images can be copied under DOS, just like any other file, making copies from within WSHFAX lets you look at the image—just to make sure you've got the right one. You can test the function by

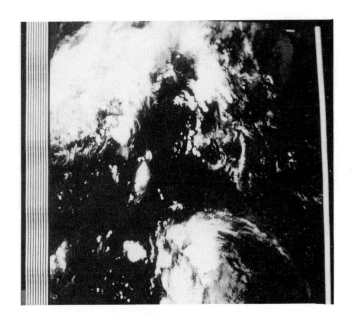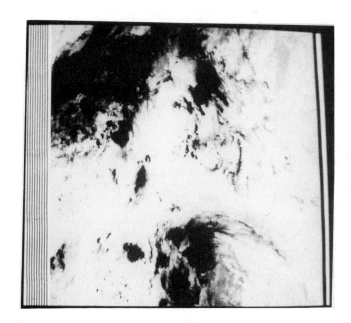

Figure 9.5—At the left (A), a METEOR IR (evening) pass. In the Russian IR format, cold objects (like clouds) appear black while warm objects are white. In effect, METEOR IR data have the appearance of a photographic negative when displayed with the normal palette setting used for image display. At the right (B) , the same image after the use of the WSHFAX <C>omplement function.

keying <S>. The upper 25% of the screen will be cleared and you will be prompted to provide a file-name for the save (*no* extension). or a <Q> to exit without a save. For the sake of experiment, type:

test <ENTER>

and the image will be saved as file TEST.480. Once the save is complete, the upper 25% of the image will reappear. At this point, key <Q> to return to the Main Menu. Key <L> to call up the load option. Now your TEST.480 file should appear in the directory. Type:

test <ENTER>

and you should see the same image you saved!

<J>et-Print

The program will support hard-copy print-out using a Laser Jet or DeskJet compatible printer. The routine is slow but produces a very high-quality image (see Figure 9.1). Since you can print file images at any time, you can delay printing sessions until a time when you don't have much else to do. You can even operate the program on another computer connected to the ap-propriate printer—if you don't have a suitable printer of your own.

This routine assumes the printer is connected to LPT1. If your computer has only a single parallel port, you will either have to employ an A/B switch box to handle both the interface and the printer, or you will have to swap cables. If you have a second printer port, connect the printer to LPT1 and the interface to LPT2. *Do not* call this routine without a printer or you will get a time-out error that will dump you back to DOS. No harm will be done (you can simply reboot), but it is aggravating.

If you do have a printer connected, keying <J> will initiate the routine. The screen will go white and the picture will be shifted to occupy the center of the screen and printing will commence. When printing is complete, you will be returned to the Main Menu and the print will be ejected. The resulting image will be about seven inches square and will demonstrate good image detail and grayscale.

<C>omplement

Keying <C> will convert the screen image to the equivalent of a photographic negative. This will allow you to view Russian polar-orbit IR images in the more familiar warm=black, cold=white format (Figure 9.5B). You can try this as often as you want because it will not permanently alter the image file.

<N>ormal

If you have keyed <C> to compliment an image, keying <N> will restore the normal grayscale palette (Figure 9.5A). This routine is called automatically

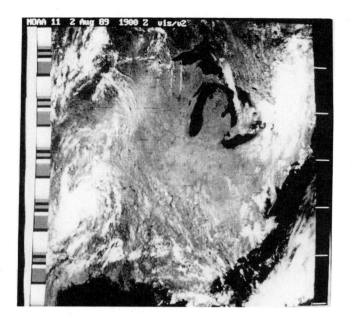

Figure 9.6—The <Z>oom option from the display menu lets you enlarge any specific area of a picture on the screen. At the left (A) you see a NOAA visible-light pass that has been contrast-enhanced via software. At the right (B) you see a subset of the same image, expanded to full-screen size using the <Z>oom function.

when you exit from the FILE DISPLAY screen so you are not left with a complemented palette!

<L>abel

The file images provided with the program are probably labeled with a single line (or partial line) of text at the top of the screen. These labels were created with this routine. If you key <L> the top 25% of the screen will clear and you will be prompted to enter a label line of up to 60 characters, or <Q> to quit without labeling. Labels can involve any legitimate keyboard characters, including spaces. Given the 60-character limit for labels, you have ample opportunity to record the spacecraft, date, time, and any other relevant data. If you enter label text (ending with **<ENTER>**), the top 25% of the image will be restored, but with the label in place. If you enter label data for an image that already has a label, the new text will over-write the old. Labeling an image does not alter the original file, unless you later <S>ave the image under the same filename. This label function allows information about the image to be saved without the complexity of additional data files.

<Z>oom

Zoom is a function that lets you fill the display screen using a subset of the file image. If you key <Z>, the "zoom" area will be defined in the upper left of the display area. If you hit the **<ENTER>** key, the defined area will shift to the right. If you continue to key **<ENTER>**, you will note that there are a total of nine possible overlapping areas where you can use the zoom function. If you want to get out of the zoom function without actually executing the routine, simply key <Q> and you will be returned to the FILE DISPLAY screen. When you have a zoom area defined that looks interesting, simply key <Z> again.

At that point, the defined area of the screen will be cleared and then the entire display area will be filled with the image data from that part of the screen. The entire display area will now be filled with approximately 25% of the original image. This new image can be labeled and saved, or you can zoom on a subset of the new image. Figure 9.6A shows an original file image while Figure 9.7B shows the result of a single zoom operation.

It is important to realize that the zoom function simply magnifies the existing detail in a file image. Because of the fine resolution of the original image, one level of zoom will typically result in a sharp image that will make it easier to pick out fine detail. You can continue to zoom successively on the image, obtaining any degree of magnification you desire. As you do so, however, the digital nature of the image will become progressively evident. There are situations where this degree of magnification may be useful. Figure 9.6C shows one such extreme.

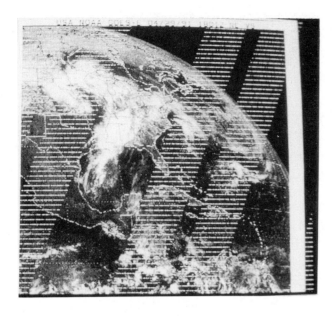 

Figure 9.7—At the left (A), a sample of a GOES WEFAX visible-light quad (NE in this case), marred by the regular streaks caused by the drop-out of WEFAX data when the VISSR instrument is scanning the earth disc. At the right (B), the same frame after the use of the WSHFAX <V>ISSR Repair function. This routine makes it entirely practical to collect high-quality WEFAX data from the primary operational spacecraft, all of which now operate simultaneous VISSR and WEFAX downlinks.

<V>ISSR Repair

When the decision was made to begin simultaneous VISSR and WEFAX operations on GOES spacecraft, it was not greeted with enthusiasm by WEFAX operators. When the spacecraft is transmitting its high-resolution data stream during the short earth-scan interval, the WEFAX signal drops out completely! It doesn't matter how big your antenna is. The WEFAX signal simply isn't there!

The result (Figure 9.7A) is the presence of unsightly streaks that mar an otherwise excellent image. Because of the relationship of the spacecraft spin rate (and hence the timing of the earth-scan intervals) and the WEFAX line rate, the short noise intervals end up as a series of vertical "stacks." The precise angle of the stacks to the vertical is constant in any image, but varies over time because of variations in the spacecraft spin-rate.

I must admit to a bit of pride about the VISSR Repair function, because it does a superb job of completely eliminating the effects of the VISSR transmission pulses! When you key <V>, a small crosshair cursor appears at the top of the screen. Pick out a stack where both the top and bottom are within the image area and use the <U>p, <D>own, <L>eft, and <R>ight keys to move the cursor until it is located at the beginning (left end) of the top line of a stack. Press <S>tart.

The cursor will reappear at the bottom of the image. Now use the <L>eft and <R>ight keys, as required, to move the cursor so it is lined up with the left edge of the bottom of the stack previously marked. Now press <E>nd. Almost by magic, the entire stack of VISSR pulses will disappear and the cursor will reappear at the top of the screen.

Since marking the first stack defined the angle of the stack, you can now move along and mark just the top line of the remaining stacks. Just key <S> when the cursor is precisely positioned. Each time you do so, the stack will disappear. When all the stacks have been repaired, key <Q> to exit the repair function and you will be returned to the FILE DISPLAY menu.

Figure 9.7B shows an example of a repaired image. After you are familiar with the routine, it takes just one or two minutes to completely clean-up such a picture. Once you are back to the FILE DISPLAY menu, you can save the repaired image to the original filename or as a new file.

<E>xport

The File Image format used by WSHFAX is unique to the program, yet there are innumerable graphics and paint programs available that can customize images saved in standard formats. <E>xport is an option that allows you to convert the currently displayed

File Image to a .PCX file. PCX files can be directly imported to paint programs such as *PC Paintbrush*, where you can crop, colorize, label features, or perform other functions. With a program such as *Graphics Workshop*, you can also convert PCX files to a wide range of other standard graphics file formats. You can also manipulate the palette, crop, scale, rotate . . . you name it. The <E>xport option is not fast, but it creates a flawless PCX image file that can greatly expand the range of things you can do with your images.

If you key <E>, the upper 25% of the screen will clear and you will be prompted to provide a filename. Enter a filename and the upper segment of the image will be replaced and file conversion will commence. As the image is converted, the pixels will be blanked on the display to provide visual feedback on how far the program has gotten in the conversion process. When conversion is finished, you will be returned to the Main Menu.

As you can see, the FILE DISPLAY menu provides a powerful range of options. File images are the most compact way to store images, but they are essentially digital copies of the display screen, with pixels coded to 16 grayscale values. As such, they don't permit extensive modification of the contrast or dynamic range of the image. Some idea of the options available with the primary satellite image data can be obtained by investigating the IMAGE DISPLAY options.

WSHFAX IMAGE DISPLAY

When you key one of the imaging modes to display a new image or use the <B>inary File Load option from the Main Menu, you will be routed the WSHFAX IMAGE DISPLAY screen. As in the case of the FILE DISPLAY screen, your options are listed down the left-hand side.

<S>creen Save

This option permits you to save the present image in the compact File Image format for long-term storage. GOES, METEOSAT, GMS, and HF WEFAX images are already processed prior to transmission and are likely to be at optimum quality when initially displayed. You could use <S>creen Save to create File Image formats immediately for these pictures. In contrast, polar-orbit images from the TIROS/NOAA and METEOR spacecraft benefit from processing prior to saving the image in the File Image format. When you do key <S>, you will be prompted to provide a filename (no extension), at which point the File Image will be created. If you precede the filename with DOS path entries, the image can be saved to another directory or drive.

<B>inary Save

This option, invoked by keying <B>, will normally be the first thing you do after image display is com-

plete. The image will be saved very quickly in the Binary File format. It's saved so quickly that you can store a WEFAX image and return to the active display in about 30 seconds! The display is not altered by a <B>inary Save operation and you can then proceed to use other IMAGE DISPLAY options, if desired. You may be puzzled by the fact that the program doesn't prompt you for a filename, but keep in mind that the <B>inary Save option is designed for speed and convenience. The WSHFAX program creates a name for the binary file based on the *time* that you save the image. If you save the file at 10:22, the filename will be 1022!

To optimize speed, binary images are saved as four 64k files, with extensions ranging from A through D. For example, a file saved at 10:22 will actually consist of the following four files:

1022.A
1022.B
1022.C
1022.D

Even though you do not label binary files, the combination of the time and the DOS date stamp allows you to identify the spacecraft—using information from your tracking program! Binary files are intended primarily for short-term storage, but you may wish to save a few for demonstration purposes. In that case, you can give the files a more useful name under DOS. Let's say you wanted to rename 1022 to something like NOAA_IR1. To do so from DOS, you could simply type:

rename 1022.* NOAA_IR1.* **<ENTER>**

and all four files would be given the new name. You could also copy the block of files in a similar manner:

copy 1022.* NOAA_IR1.* **<ENTER>**

By using DOS path entries prior to the new name, the files could be copied to another directory or drive. When you want to delete the files, simply type:

erase 1022.*
OR delete 1022.* **<ENTER>**

The <U>nattended Operation option from the Main Menu saves images in the Binary File format, so you will be using these files extensively.

<I>nvert

In the case of south to north polar-orbit passes, the image will be displayed "upside-down," in the sense that north will be at the bottom of the image. WEFAX charts are also transmitted inverted. In either case, keying <I> will solve the problem. The program requires a few seconds to perform a complete inversion of the data in the image buffer, at which point the image will be redisplayed in its inverted form. You can try this option to see the effect. To redisplay the image as it was originally, simply key <I> again!

<C>omplement RAM

This option complements the image data in the RAM buffer.

Processing Options

If you have examined the VIS_RAW and IR_RAW Binary Files, you will note that they are somewhat less than optimum in terms of display quality. The VIS_RAW file looks too dark while the IR_RAW file seems too light. This is perfectly normal if you've set the SAT CONTRAST control properly. (It is an accurate reflection of how the image was transmitted.) The *Processing Options* represents a set of image enhancement routines that let you tailor the file data to get precisely the image you want. I will defer discussion of these options until Chapter 10, because they will make more sense after you have a basic introduction to digital image processing.

<U>nattended Operation

This option lets you capture images to disk while you are away. I will discuss this option in greater detail once you are familiar with the basic IMAGING OPTIONS.

IMAGING OPTIONS

All the various imaging options share a few features in common. First, when you key one of the imaging modes, you will be routed to the IMAGE DISPLAY screen. You should be familiar with this display by virtue of playing with the Binary Files. Normally, the last entry on the menu options list is **COMMAND**. This indicates the *command mode* in the sense that the system is ready to receive input from the keyboard. When you call an imaging option, however, the **COMMAND** entry will change to **ACTIVE**. This tells you that the system is in an *active* imaging mode.

AM/FM Switch Settings

All direct-broadcast satellite formats use AM subcarrier modulation. The output of the satellite receiver should be connected to the SAT input jack (J1) and the AM/FM switch should be in the AM position. All HF WEFAX transmissions use FM. The output of your short-wave receiver should connect to the HF input jack (J2) and the AM/FM switch should be in the FM position. In short, use AM for all modes *except* HF WEFAX, where you should use the FM position.

<A>borting Display

Most of the imaging modes will cause the system to return to the command configuration when an image is complete, but there may be situations where you wish to abort image display. If the screen shows an **ACTIVE** prompt and you wish to abort, simply hit the <A> key and the **ACTIVE** prompt will be replaced by **COMMAND**. This is the *only* single-key entry in the program that *is* case-sensitive. You must key a *lower-case* <A> in order to abort. This will occur automatically if the <CAPS LOCK> is not engaged, or you don't use the <SHIFT> key. If <CAPS LOCK> is engaged, the program will ignore your keyboard input and continue the display.

Automatic Display Modes

There are two types of imaging modes: automatic (or *autophasing*) and manual. WEFAX, VIS NOAA, IR NOAA, and METEOR satellites fall into the autophasing category. In the case of WEFAX, image display begins when a valid start tone has been detected, autophasing is complete, and a delay has been executed to bypass most of the phasing interval. In the case of VIS NOAA, IR, NOAA, and METEOR, display begins almost immediately, once the proper sync pulse has been identified.

As in all imaging modes, you can abort display at any time by keying <A>. If you don't, the image will proceed to completion. The NOAA and METEOR routines will return you to the **COMMAND** mode when display is complete. In contrast, WEFAX will simply wait for a new start tone and display the next image. This can go on indefinitely, allowing you to display sequential WEFAX images for weeks on end! To save an image, you will have to key <A> at the end of the image to return to the **COMMAND** mode.

To use autophasing, the following requirements must be met:

1. The image signal must be relatively noise-free. Excessive noise on the signal can lead to false starts in the case of WEFAX and can result in misphasing in any of the modes. Noise is not usually a problem with WEFAX, but is a fact of life at the early stages of a NOAA or METEOR pass. For best results, wait until the signal has reached full-quieting before selecting VIS NOAA, IR NOAA, or METEOR display.

2. The interface SAT CONTRAST control must be properly adjusted, as outlined in the earlier discussion of the Main Menu <O>scilloscope function. Detecting sync and phasing pulses involves intricate operations with respect to signal levels. If SAT CONTRAST is way out of adjustment, the routines may not operate reliably.

<W>EFAX

When you key <W>, the system will shift to **ACTIVE** status and begin looking for a valid WEFAX start tone. Nothing will happen, even if you start during a picture transmission, until a start tone is detected. Even then there will be a delay, bypassing the

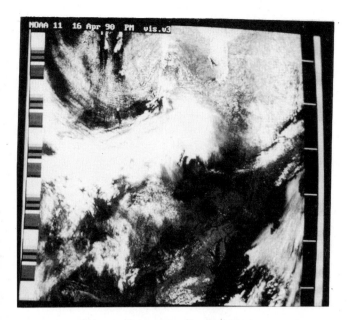

Figure 9.8—Visible-light data from a NOAA 11 afternoon ascending pass, acquired using the WSHFAX <V>IS NOAA mode. The image has been inverted to place north at the top of the display. Ice-covered James Bay is visible at the top of the image.

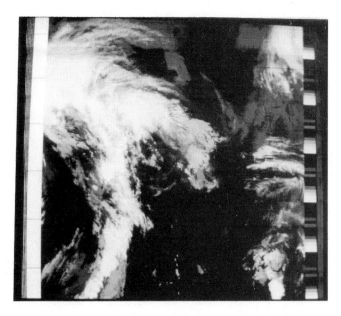

Figure 9.9—Infra-red (IR) data from a NOAA 12 morning pass. The WSHFAX <I>R NOAA mode was used to display the image. The pass was taped using an 8-mm video tape deck, permitting recovery of both IR and visible-light data. The raw video data for both formats, displayed side-by-side, can be seen in Figure 9.11. The display here was enhanced to improve the dynamic range of the IR data using processing options in the WSHFAX program that are described in Chapter 10.

start tone and most of the phasing interval. Do not expect the display to begin until almost 25 seconds after the beginning of the start tone! For versions of this program used in countries where METEOSAT or GMS is the active WEFAX source, the timing will be a bit different. These spacecraft transmit images with only five seconds of phasing, compared with the 20 seconds used by the GOES spacecraft. In the case of programs shipped to Europe, Africa, and Asia, display will begin approximately 10 seconds after the beginning of the start tone.

<V>IS NOAA

Keying <V> will result in full-screen display of the visible light channel from a TIROS/NOAA pass (Figure 9.8). Assuming the signal is essentially noise free, and that you've properly adjusted SAT CONTRAST, display will typically begin within a few seconds of pressing the <V> key. About six minutes of the pass will be displayed.

<I>R NOAA

Keying <I> works identically to the <V>IS NOAA option except that the IR channel data are displayed (Figure 9.9). During night passes, where both satellite downlink channels contain IR data, the routine will select one or the other for display. There is no control over which channel is selected. If you must display one

specific IR channel for evening passes, I would suggest that you use the <A>PT option under MANUAL IMAGING.

<M>ETEOR

Keying <M> will initiate display of standard 120 LPM METEOR images (Figure 9.10). In daylight these will represent visible-light data. At night, IR data will be transmitted. In the latter case, you may wish to use the <C>omplement function when display is complete and you are returned to the **COMMAND** mode.

Manual Display Modes

The APT, 240 LPM, 120 LPM, and HF WEFAX options all use manual phasing. When you enter the mode, the image will begin to read out from the top of the display area. Because the start of the image is random, it will almost certainly be misphased. The left edge of the image format will be somewhere to the right of the left edge of the display frame. You initiate phasing by hitting the <P> key, at which point the image will "step" to the left in small increments as it reads out. When the left edge of the image (typically

Figure 9.10—Daylight (visible) Russian METEOR pass, displayed using the WSHFAX <M>ETEOR mode. This was a south-north pass and was inverted using the display inversion routine so that north is at the top of the display. This causes the characteristic METEOR sync pulse to appear on the right side of the display instead of the left. The label was added to the image using the image label routine in WSHFAX.

a sync pulse) reaches the left edge of the display frame, key <P> to stop phasing. At that point the image display will reset to the top of the screen and read-out will begin with the image in the proper phase relationship.

Although the image has started to read out as soon as you enter a manual mode, this initial display is only to assist you in phasing the image. No data are added to the RAM buffer at this point and, if you allow the image to continue in this manner, it will simply reset when the bottom of the screen is reached and then begin to overwrite earlier data. You *must* phase the image to start valid image display!

When the displayed data has filled the entire screen (about three minutes for 240 LPM, six minutes for APT and 120 LPM, and 12 minutes for HF WEFAX), the system will automatically return to the **COMMAND** state. Now you can manipulate the image, save it, or perform any other function available from the IMAGE DISPLAY options.

<A>PT

This mode is a manual phasing mode designed to display *either* the visible light or IR TIROS/NOAA image format. You can initiate display any time the signal has become relatively noise-free by simply key-

ing <A>. When display begins, you can key <P> at any time to begin phasing.

Phasing in this mode is your means of selecting either visible light or IR display. To select visible data, wait until the sync pulse—followed by the *black* pre-earth scan (see Chapter 4)—moves over to the left margin of the display area. Key <P> again to stop phasing. At this point the image will reset to the top of the screen and visible-light display will start. To select IR, wait until the sync pulse—followed by the *white* pre-earth scan—is aligned with the left edge. Press <P>, and phased display of the IR data will begin.

At night, both channels will consist of IR data and you can select either one for display, depending upon how you phase the image. Recording a pass will let you display both channels of data, since you can replay the tape and phase on the alternate channel if desired. If you wish to display visible and IR data side by side, use the <1>20 LPM mode discussed in a later section. When display is complete (about six minutes), you will be returned to the **COMMAND** option, at which point you can save or manipulate the image. I suggest that you get in the habit of doing a <B>inary Save as your first step. This allows you to recover the original image data, no matter what other procedures you have used.

You can abort display at any time by hitting the <A> key. If you make an error in phasing, simply hit <A>. When **COMMAND** appears, go to the Main Menu with <Q> and resume display by hitting key <A>.

<2>40 LPM

This routine will probably get less use than any other in the program It can be used to manually display WEFAX images, but the automatic WEFAX mode is so reliable, you will only need <2>40 LPM when you are trying to display a *very* noisy picture. It can also be used if you are lucky enough to encounter a 240 LPM COSMOS spacecraft. When you key <2>, display will begin. You can use <P> to start phasing, using it again when the image has slipped into the proper phase relationship. The image will then reset to the top of the screen and begin to read-out. Display will require about three minutes, at which point you will be returned to the **COMMAND** mode.

If you are using this mode for manual WEFAX, key <2> when you hear the start tone. Begin phasing about half-way through the 20-second phasing interval. COSMOS images can be displayed whenever the signal achieves a noise-free level. Typical 240 LPM COSMOS signals mark the left edge of the image with a column of numbers. Align on the left edge of the display when phasing.

<1>20 LPM

This mode operates in a similar fashion to the MANUAL APT mode, but it's designed primarily for

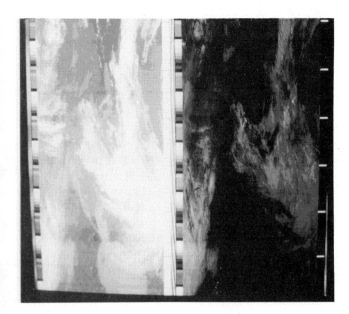

Figure 9.11—The WSHFAX <1>20 LPM mode can be used to provide side-by-side display of visible and IR data for TIROS/NOAA passes. In this case, the videotaped pass used for Figures 9.8 and 9.9 was replayed to illustrate this feature. This is raw image data, so the visible channel image (left) looks quite dark while the IR image (right) appears very light. Chapter 10 illustrates how different contrast enhancement routines can be applied to the different components of the composite image.

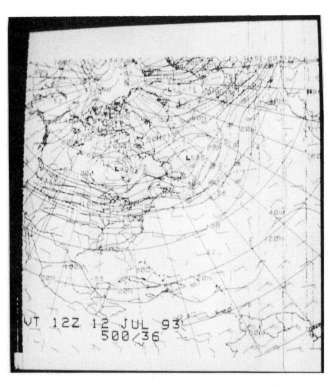

Figure 9.12—A sample of a typical weather chart, displayed using the <H>F WEFAX mode in the WSHFAX program. A truly ancient short-wave receiver, barely stable enough to hold the proper tuning for the duration of the picture, was used to receive NAM (Norfolk, VA) on 8.080 MHz. While standard VGA resolution is marginal for these ultra-high resolution charts, you can still derive useful information from them. Such charts are transmitted inverted, so the WSHFAX inversion routine was used to place north at the top of the display.

display of 120-LPM METEOR imagery. When display begins, the distinctive METEOR sync pulse (see Chapter 4) will be located somewhere within the length of the horizontal display. Hit the <P> key at any time to begin phasing, at which point the pulse train will step toward the left. When it reaches the left margin of the display, key <P> again to stop phasing. The image will reset to the top of the screen and begin to read out in proper phase.

When the display is complete, you will be returned to the **COMMAND** mode. You can abort display at any time by using the <A> key. Again, I suggest <B>inary Save as your first option so you can recover the original image data later, if required.

Daylight METEOR images are always visible-light data, switching to IR at night. With an IR image, you may wish to use the <C>omplement function to view the image in the more familiar TIROS/NOAA IR format where warm objects appear dark and cool objects light.

The <1>20 LPM mode can also be used to display TIROS/NOAA visible and IR data in a side-by-side format (Figure 9.11). Use the phase function to position either sync pulse at the left edge of the display.

<H>F WEFAX

Verify that the AM/FM switch is in the FM position and begin display by keying <H> when you hear the start of the phasing interval. Although <H>F WEFAX is listed as a manual mode, it is actually a hybrid. It does not autostart (like WEFAX), but when you key <H>, it will phase automatically and begin display. You *must* key the mode during the phasing interval. If you start during the start time or after image transmission has begun, the image will probably be misphased.

Once display begins, you can go get a cup of coffee. Reception of the entire image will require about 12 minutes! You can abort display at any time using the <A> key. Otherwise, the system will return to

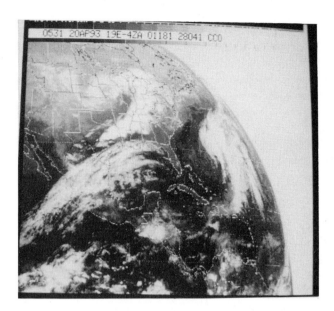

Figure 9.13—Using the <H>F WEFAX mode lets the WSHFAX program display GOES images transmitted on HF. Assuming reasonable propagation, the image quality can be as good as that achieved using S-band WEFAX. This image was received from NAM (Norfolk, VA) on 8.080 MHz.

COMMAND status when the transmission is complete. Do a <B>inary Save for insurance purposes, and then proceed to play with the image with any of the DISPLAY MENU functions. While the resolution of the standard VGA screen is marginal for HF weather charts (Figure 9.12), the results with grayscale imagery (Figure 9.13) are quite impressive.

<U>nattended

Keying <U> will invoke a routine that permits you to log polar-orbit passes or WEFAX transmissions when you aren't at home! In the case of polar-orbit spacecraft, this is easy to implement because it is practical to use scanning receivers and omnidirectional antennas. In order to use this option, you must prepare a "script" file (AUTO.DAT) that contains certain basic information about the spacecraft you wish to receive.

AUTO.DAT must be an ASCII file (no "high bit" control codes). You can prepare it using almost any word processor in the nondocument or ASCII mode. Listed below are the contents of a sample AUTO.DAT file:

```
07-09-1993
20:05 VIS NOAA
07-09-1993
```

21:47
VIS NOAA
END

The data for individual passes are in a 3-line format, which you can see by looking at the first three lines of this sample file. The first line is the date of the transmission we want to log. It should be entered in precisely the format used by your computer when displaying the date. This will typically be MM-DD-YYYY or MM/DD/YYYY. Other ordering patterns may be used in Europe and elsewhere. The format your computer uses when it presents you with a response to the DATE command under DOS is what you should use.

The second line contains the time (HH:MM) when we want to begin display. This particular pass is a NOAA 11 pass that WSHTRAK told me would have an AOS at 20:01 hours and an LOS at 21:15. Since WSHTRAK calculates AOS and LOS as if the pass were overhead, the interval between the two is always 14 minutes. If you want to assure the best possible signal—assuming we will display six minutes of pass data—we want to initiate display about three minutes before overhead (or the highest point in the pass). The high point of the pass will always be *approximately* seven minutes after AOS, as computed by WSHTRAK. I simply add four minutes to the AOS time to derive the "turn on" time for acquisition—in this case, 21:05. Since this particular pass was part of a straddling pair and well off to the east, this was a very conservative time choice. In the case of a near overhead pass, you could bias the time several minutes earlier or later and still be assured of a reliable signal. In most cases however, I stick with my "4-minute" rule.

In the case of a WEFAX transmission, it is quite possible that the desired image may be sent a bit late, but hardly ever will it arrive early. For a WEFAX image, I would set the time about one minute early—not enough time to tangle with the previous image, but early enough to assure that I don't miss the start tone of the desired image.

The third line of the "script" contains the mode, in precisely the format used on the Main Menu of the WSHFAX program. There are four modes that are recognized:

WEFAX
VIS NOAA
IR NOAA
METEOR

Note that a space is included for the NOAA options. You'll also note that these correspond to the four autophase modes on the Main Menu. You cannot select a manual mode because there will be no one around to phase the image!

The simple 3-line script thus tells the computer the day, the precise time to start display, and the display mode to use. You can have as many as 25 of these 3-line entries, each representing an image to be acquired, in your AUTO.DAT file. There are a few simple rules that must be followed:

1. Entries must be sequentially arranged by date, first to last.
2. For any given day, the entries must be sequential with respect to time.

Our sample listing has only two entries. The last line—END—is just as important as the others, because it tells the computer that the file is done. The word END must be on the last line of the file and it should be in uppercase.

When I sit down to make up my AUTO.DAT file, I will typically compile a sequence of "hot" passes for the next week or two. With a system like this, you will be getting a *very* large number of pictures, so you can afford to be selective!

After you've completed your own AUTO.DAT file, you can run WSHFAX and key <U> from the Main Menu to invoke the <U>nattended mode. Be sure, however, that the DOS DATE and TIME functions reflect the GMT/UTC time and date. Also, check to make certain that you've set your SAT CONTRAST control using the <O>scilloscope function. Your receiver should be scanning, or set to the frequency required for the first spacecraft in your file.

Nothing spectacular will happen when you key <U>. The screen will go blank and you will be tempted to suspect that the entire system has died! This is far from the truth. At precisely the time indicated by the file entry, the screen will pop back to life in the IMAGE DISPLAY mode and the incoming image will be displayed, just as if you were doing all the work! When display is complete, the hard drive will cluck to itself for a few seconds as the Binary File is created. Then the screen will go blank and the system will wait for the next pass.

You can wake your sleeping beauty by hitting any key on the keyboard, returning you to the Main Menu. One of my rituals, after dinner, is to go down, wake up the system, and use the <B>inary File Load function to see what spacecraft I caught that day. Once I've played with the pictures, processing and saving File Images, I put it back in the <U>nattended mode. Here is where the neat part comes in! The computer will check the first entry in AUTO.DAT. If the date or time indicated for that entry is out of date (prior to the current date and time), it will skip it and look for the next entry that has yet to occur! That's why it's practical to build an AUTO.DAT file that may stretch several weeks into the future. When I get time, I fire up WSHTRAK and my word processor, editing out the old entries and adding new ones. With the <U>nattended function, the computer is always ready to justify itself, even if I'm not there to pound the keyboard!

Well, that covers most of the features of WSHFAX, *except* for the image-processing functions. These will be discussed in Chapter 10. Digital image processing looks like the technological equivalent of magic, so be prepared for some surprises when we address the topic. You can expect additional features in releases of WSHFAX available by the time this edition is in print.

Chapter 10

Working with Digital Images

INTRODUCTION

Two adages come to mind as an introduction to this chapter: "a picture is worth a thousand words" and "seeing is believing." These days, neither one has validity. The pictures captured and stored by WSHFAX, while modest by today's digital image standards, are made up of almost 250,000 bytes—equivalent to about half the text of this edition. Today's digital pictures are worth a great deal more than 1,000 words!

Once you start to view images in terms of a huge collection of numbers, they also tend to lose any sense of concrete reality. Numbers can be manipulated or altered and, if they are, so are the pictures they represent. Seeing is no longer believing. Like the national debt, the numbers can be manipulated to represent just about anything. This is a major boon to those of us working with satellite images, because we can process our pictures in ways that were impossible even a decade ago!

When picture displays were purely analog, there was comparatively little one could do to improve the quality of a poor image, emphasize important features, or remove the effects of noise or interference. You could alter the gain levels of the display equipment, employ nonlinear circuit elements, and play with the image in the darkroom. All of this could help, but it was all so time-consuming that small ground stations would only rarely make attempts to significantly manipulate images. Given the nonlinear response of fax papers, the limited dynamic range of CRT phosphors, and a host of other practical problems, it was enough to strive for images that looked more-or-less right!

As Jeff Wallach has outlined in his discussion of HRPT in Chapter 7, today's spacecraft excel in the precision of high-resolution digital output. While the APT and WEFAX downlink signals arrive via an analog link, almost every aspect of the signals are more stable and reproducible than was the case in the pioneering days of amateur weather-satellite stations. When we convert our image back to digital form, we have tremendous opportunities to process and enhance the pictures. Un-

like analog technology, where you basically had one chance to get it right, we now have digital copies of the original data stream safely stored on disk. We can perform all sorts of experiments on copies of these digital files, without compromising the original data.

There is no way to provide a comprehensive introduction to digital image processing in a single chapter. The best I can do is to hit some of the high points and give you a start in performing some of your own experiments. An excellent reference that treats the subject in greater depth is G. R. Braxes' *Digital Image Processing* (Prentice-Hall, Inc., 1984, Englewood Cliffs, NJ 07632, 182 pp.). In this chapter, I emphasize some basic processes, many implemented with BASIC code. BASIC *is* slow, but it is also easy and reliable to use in your first experiments with manipulating images. Even now, I tend to test new routines using short BASIC programs similar to the one I present here. When they are thoroughly debugged, I implement them in the faster Assembly Language module.

Digital image processing is a fascinating subject and provides endless opportunities for experimentation. It doesn't cost anything and the WSHFAX program generates all the image files you will ever need to play with the pictures. This is one area where you don't have to write a complete imaging program to perform very useful operations.

HANDLING THE BINARY FILE FORMAT

The Binary File format is your most flexible option because image data are coded in a simple format. The pixels are stored in a linear sequence, starting with file .A and ending with file .D. Each pixel takes up a single byte, so there is no complicated formatting to worry about. Figure 10.1 is a simple BASIC listing that has provisions for three different functions:

1. Moving data from the four Binary Files for the image into RAM where the data are easier to access.

2. Code for moving the binary data in RAM to the screen for display

3. Space for code to transform the data in going from RAM to the screen display.

```basic
'========================================
'define screen and set palette
'========================================
        SCREEN 12
        CLS
        r = 0
        FOR n = 0 TO 60 STEP 4
        c = INT((65536 * n) + (256 * n) + n)
        PALETTE r, c
        r = r + 1
        NEXT n
        p = 128: 'pixel start
        l = 0: 'line start
'========================================
'identify and load binary file
'========================================
        INPUT "Binary file to load (no extension)..."; f$: CLS
        DEF SEG = &H6000
        file$ = f$ + ".a"
        GOSUB sndscrn
        file$ = f$ + ".b"
        GOSUB sndscrn
        file$ = f$ + ".c"
        GOSUB sndscrn
        file$ = f$ + ".d"
        GOSUB sndscrn
        DEF SEG
'========================================
'post-display functions can be inserted here
'========================================
hold:
        GOTO hold
'========================================
'moves 64K of segment 6000h to the screen, up-dating
'line and pixels values in the process
'========================================
sndscrn:
        BLOAD file$, 0
        FOR ram = 0 TO 65535:
                v = PEEK(ram): 'get ram value
        '========================================
        'insert video transform here
        '========================================
                PSET (p, l), v: ' write pixel
                p = p + 1

                IF p = 640 THEN p = 128: l = l + 1
                NEXT ram
                RETURN
```

Figure 10.1—A BASIC program listing for transferring a WSHFAX binary image file to the standard 640 × 480 × 16 VGA screen.

The listing in Figure 10.1 takes care of no. 1 and no. 2. Much of what we will discuss in this chapter represents options for no. 3. If you insert a single line below the video transform notice in the *sndscrn* subroutine:

$$v = INT(v/16)$$

you can test the code listing. This single line performs a linear transposition from the 0 to 255 values stored in RAM to the 0 to 15 values required for screen display. When you run the program, you will be prompted to enter the name of a Binary File (no extension). If you are running your BASIC code from a different directory, you must include the path data with the name. For example, to load the IR1_RAW file in your \SAT directory, type:

\sat\ir1_raw <ENTER>

Assuming you have no typos (the program is included on the WSH Program disk as FIG10_1.BAS) in your listing, the screen will clear and the image will begin writing to the screen. It you are patient, you will eventually see an image identical to the one you saw when you used the <B>inary File Load function from WSHFAX. When image display is complete, the image will remain on the screen until you hit <Ctrl-Break>. BASIC may be slow, but it does have the advantage that you can use <Ctrl-Break> as soon as you see that your code is working (or not working!) as you would like.

The rest of this chapter will be devoted to a combination of theoretical discussion, examples drawn from the use of the WSHFAX program, and, where feasible, examples where you insert your own video transform functions in this program.

NOISE REDUCTION

There are two major sources of noise that impact our pictures. We have to contend with *random* noise resulting from the strength of the received signal falling too low with respect to the noise generated by the receiving system itself. This always occurs at the beginning and end of a pass. It can also occur if the spacecraft moves through a *null*, or lower-gain portion, of the antenna pattern—even when the spacecraft is higher in the sky. Noise can also be a problem with passes off to the east or west of your station. Since the spacecraft never reaches a very high elevation, its signal is often quite noisy. Natural RF noise sources such as the sun and static from thunderstorms, can also intrude on the signal. The result is a "salt and pepper" pattern on the screen. The noise is random in two different ways. First, the position of pixels disrupted by noise cannot be predicted. They will be scattered in a random pattern on your display. Secondly, the amplitude of the noise pulse is unpredictable. As a result the gray-level of an impacted pixel is more or less random.

Noise that becomes visible on an image indicates that our received signal-to-noise ratio (SNR) is getting marginal. Curiously, this is about the only area where our simple APT installations have an advantage over the more exotic HRPT systems discussed in Chapter 7. With an adequate SNR, an HRPT receiver will deliver an essentially *perfect* reproduction of the incoming digital data stream. If the SNR becomes marginal, you won't see noise, as you do with an APT system. Instead, you won't see anything at all! In contrast, our simple APT installations will continue to display, even if the desired signal is completely obscured by noise! The subcarrier pre-filter and post-detection filters in the WSH Interface (Chapter 5) do a good job in suppressing the effect of noise *outside* the video passband. However, there is no way that they can suppress noise *within* the video passband. Thus, with a marginal SNR, some noise will come through. Fortunately, as we shall see, there are ways to significantly reduce the *visual impact* of such noise.

Nonrandom noise can also be a problem. Although the amplitudes of such noise pulses may be random, the pattern they make on the screen is not. Such noise can occur as a result of malfunctioning electrical equipment (light dimmers, for example), where the pattern is often related to the timing of the 60-Hz (or sometimes 120-Hz) ac line. You'll also encounter ignition noise from car engines, where the pattern is related to the relationship of the throttle setting of the engine and the number of cylinders. You may even see the nonrandom noise patterns generated by GOES VISSR operations, where the pattern is determined by the spacecraft spin rate and the length of the high-data-rate Earth scan. In some ways, nonrandom noise is easier to cope with than the random variety. Knowing something about the spatial pattern of the noise allows us to compensate for it without needlessly altering areas of the picture where the noise doesn't occur.

Random Noise

All digital-processing techniques for reducing the effects of noise on image display are based on a simple premise. Although we can never know the actual value of a pixel that has been corrupted by a noise pulse, we can make some inferences about it on the basis of neighboring pixels. The premise works because the magnitude of noise pulses is random, but pixel patterns are not. They reflect brightness variations that are nonrandom with respect to both shape and luminance. There are many ways to take advantage of this premise, some quite complex in mathematical terms, but the simplest approach is old-fashioned averaging.

Imagine a grid of 9 pixels on the display:

```
1 2 3
4 5 6
7 8 9
```

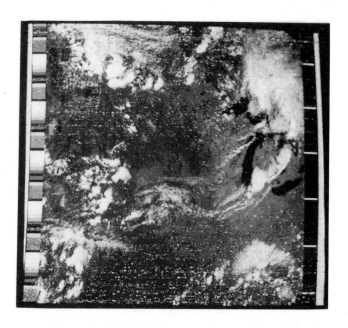

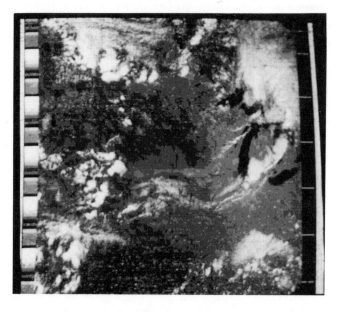

A   B

Figure 10.2—(**A**) A NOAA 11 pass off to the west of the author's location, obtained in the midst of a severe thunderstorm. The combination of low pass elevation, rain, and lightning static resulted in a moderate amount of noise on the image. (**B**) The same image after the use of the <F>ilter option from the FILE DISPLAY menu. This function employs the 4-point, "nearest neighbor" pixel-averaging routine discussed in the text. While there is a reduction in resolution, the dramatic effect of the significant reduction in noise interference can be a worthwhile trade-off in many situations.

Assume that pixel no. 5 has been corrupted by a noise pulse, so we cannot know its true value. We could, however, reconstruct that value by assigning it an *average* value based on the brightness values of its neighbors. Deciding which neighbors to use presents some interesting trade-offs, which we will discuss in a moment. First, let's deal with a few other sticky points.

Assuming we can "fix" a corrupted pixel by averaging, we need to keep in mind that we have not recovered the *true* value of the missing pixel but merely have *inferred* its value based on its nearest neighbors. Even so, which pixels do we fix? There are almost a quarter of a million pixels on a 640 × 480 VGA display. While some pixels may be obviously "out of step" with their neighbors, it is impractical to go through them manually doing a fix-up routine.

So what about using the computer to do the job? You can, but the real sticking point is developing criteria to distinguish "bad" pixels from "good" ones. For openers, you could scan the entire image, computing the average difference between any pixel and its neighbors. You would then have to determine how much a pixel would have to differ from this average to be labeled "bad" and thus subjected to a fix. If you make the differential large, you will not accidentally

fix many "good" pixels but you are more likely to miss some "bad" ones. Make the margin smaller and you will hit more of the "bad" ones, but you will alter more good ones as well. The methods range from simply making an arbitrary choice to much more elaborate statistical techniques. Unfortunately, math of this sort is far easier to implement in BASIC than it is in Assembler. Most of your experiments would take a *long* time.

There is an easier way. Change *all* the pixels by averaging with their neighbors. This approach obviously changes "good" *and* "bad" pixels. Although you hit all pixels corrupted by noise, you average out pixels that were actually distinguishing features in the original image. The result, inevitably, is loss of both spatial and tonal resolution. The more pixels we average, the greater the image degradation. If we average all eight pixels that surround any specific pixel, we have taken the most conservative approach to resetting each to its probable value. However, we have also taken the biggest hit in terms of image spatial and tonal resolution.

Of the eight pixels around any specific pixels, only four are actually the nearest neighbors—numbers 2, 4, 6, and 8 in the grid illustrated previously. Pixels 1,

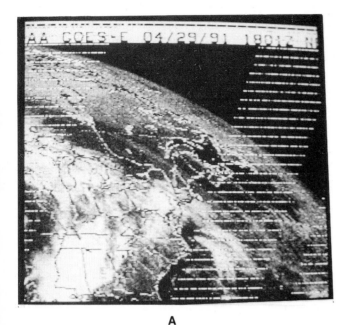

A   B

Figure 10.3—(**A**) A high-magnification view of a very small portion of a WEFAX quad, with VISSR-noise pulses, obtained using the <Z>oom function. The VISSR "streaks" are confined to single lines with a definite geometric relationship. (**B**) The same area of the image after the use of the <V>ISSR Repair function discussed in the text. The predictable relationship of damaged pixels in this case permits an almost flawless "repair" with no real loss of display resolution. The full-frame images from which these magnified views were derived can be seen in Figure 9.8 in Chapter 9.

3, 7, and 9 are in a diagonal relationship to the central pixel (5) and thus a bit farther away. If we average the four nearest neighbors and use that value to reset each pixel, we achieve a considerable degree of reduction in terms of visible noise effects while minimizing resolution losses.

This is precisely what the <F>ilter routine in WSHFAX accomplishes. Figure 10.2A shows a pass off to the west of Michigan, logged during one heck of a thunderstorm (you can see the well-defined frontal system draped across Michigan). Between the drenching rain, tremendous lightning, and the low elevation of the pass, there is significant noise throughout most of the picture. Figure 10.2B shows the same image data after the screen image has been processed by the <F>ilter routine, using nearest-neighbor averaging. The routine has an almost magical effect on noise, yet the image still retains considerable detail. You can see the difference in resolution if you compare the two versions closely (it may not be as obvious in the half-toned figures), but if the goal is to achieve a more aesthetically pleasing image, <F>ilter will do the job with relative speed.

If you like statistics, matrix math, and other exotic pursuits, there are far more elegant approaches to noise reduction. The challenge is to implement them so that they work reasonably fast. It's no accident that

folks into serious noise-processing tend to use super computers!

Nonrandom Noise

It should be obvious from the preceding discussion that the biggest problem with noise processing is not repairing the damage. It's knowing where to do the repairs! In the case of nonrandom noise, the definite spatial distribution of the noise effects makes it much easier to deal with them without having an adverse impact on the rest of the image.

An excellent case in point is the noise streaks caused by simultaneous VISSR and WEFAX transmissions from the GOES spacecraft. Figure 10.3A shows a high-magnification zoom of the limb of the Earth at a point where one of the "stacks" of VISSR-noise pulses occurs. Although the amplitudes (grayscale values) within the pulse are random, the pulse itself is clearly defined. Each pulse occupies only a single line and has a definite start and end-point.

In digital terms, repairing such a streak is simple. Each faulty pixel in the short segment of the line that defines the pulse can be replaced by the average value for the pixel above and below it! What's more, the number of lines between pulses is also constant. If we compute the angle of the "stack" (accomplished in the WSHFAX <V>ISSR Repair routine by marking the beginning of the stack at both the top and the bottom), it is a relatively simple matter to repair all the pulses that make up a single "stack." Since the angle of all other stacks is the same (the spin rate of the spacecraft will not alter appreciably within the time frame of a single transmission), other stacks can simply be repaired by marking the beginning of the pulse at the top of each stack.

The repair is nearly flawless (see Figure 10.3B). The vast majority of other pixels that make up the image are unaltered in any way. The pixels of each pulse are, of course, "fake" in the sense that they were reconstructed by averaging, but the averaging process is rational in this case. The results, in most cases, are probably quite close to what the pixels *should* have been. The matter is somewhat moot. Lacking the repair function, there would have been no useful data for those pixels in any case. Images marred with such VISSR pulses would have been permanently flawed if we were using analog techniques. I still have quite a few perfectly functional analog fax machines on hand. Now you understand why they are gathering dust!

COMPLEMENTING IMAGES

Image complementation is a process that reverses the brightness relationships in a picture. If the original image was a positive version of the picture (bright areas corresponding to high luminance in the original subject), then a complemented version of that image will be the equivalent of a photographic negative. If such a negative-image is complemented, the result is a positive image.

One of the primary uses for complementation concerns Russian METEOR infrared images. Since the inception of IR imaging on US spacecraft, the downlink signals of direct-readout spacecraft have been modulated so that cold objects are at the white end of the display dynamic range while warm objects are at the dark end of the display grayscale. Since clouds are typically cold, relative to the land or water below them, the result is an image that looks somewhat like a visible light image—white clouds against a darker background. The Russian modulation system reverses this relationship—cold objects are dark and warm objects are light.

Although this is the reverse of the system employed by US spacecraft, neither system is inherently right or wrong. In fact, the Russian system is a bit more rational in that increasing luminance implies increasing radiation. Hence, dark objects are cold (little heat radiance) and white objects are hot (high IR radiance). That aside, the images produced with that modulation format are harder to interpret if you expect clouds to be lighter than the surface below them! Complementing Russian IR imagery (compare Figures 9.5A and 9.5B in the previous chapter) creates a picture that is

easier to interpret. Still another use for complementation is that it is easier to pick out features at the light or dark end of the display dynamic range if the brightness relationships are reversed. There are two primary ways to complement an image using computer graphics display and WSHFAX uses both of them.

Complementing the Color Palette

Unlike earlier PC color displays, the VGA hardware does not hard-wire color values. The hardware of a basic VGA card can code for 262,144 discrete color values, but the limits of the hardware allow only 16 of these colors to be displayed simultaneously in the 640 × 480 mode. All colors are the result of mixing three primary colors: red, green, and blue. Each of these colors can have 64 possible intensity values, ranging from black to full-intensity brightness for the particular color. The number of possible color combinations is thus 64^3 or 262,144.

The color possibilities of the VGA hardware are implemented in BASIC using the PALETTE statement. PALETTE has the following syntax:

$$\text{PALETTE r, c}$$

In this format, r corresponds to the video value (ranging from 0 to 15 in the 640 × 480 standard VGA mode) that we write to a pixel. In effect, r represents one of the 16 available color slots. Unlike hard-wired displays, however, these 16 slots don't have a dedicated color. What you get when you write a value between 0 and 15 to the screen depends on the color code (c above) that is matched with that particular value of r when the PALETTE statement was executed. The numerical value of c defines the color as a result of the sum of the three primary color components:

$$c = (65536 * blue) + (256 * green) + red$$

where *blue*, *green* and *red* are the desired color intensity (range of 0 to 63) for each of the primary colors. If the values for *blue*, *green* and *red* are equal, then c will represent a shade of gray from 0 (black) to 63 (maximum white).

When WSHFAX boots, it sets up the 16-step grayscale palette as follows:

```
r = 0
for n = 0 to 60 step 4
c = (65536 * n) + (256 * n) + n
PALETTE r, c
r = r + 1
next n
```

Since n is stepped through the range of 0 to 60 in equal steps—while r is incremented by one with each of the 16 steps—the palette is set up with equal-step grayscale values for r from 0 to 15. (They are grayscale values because the same weighting, n, is applied to each color component.) The result is that pixel values between 0 and 15 cause the posting of equal-step grayscale values from black to white.

To complement the image, all we have to do is reverse the order in which the values of c are assigned to r. Instead of starting with $r = 0$ in the first line, we set r to 15 ($r = 15$). Then, in line 5 (above), we decrement r with each loop ($r = r - 1$) instead of incrementing it. Try this in the listing in Figure 10.1 and see the result on the images you display. In the very short time that it takes to execute the routine, all palette assignments are reversed and the screen will switch instantly to its complement values. In switching the screen back to the "normal" palette, we simply need to execute the routine as originally written. This is precisely the approach taken to implement the <C>omplement and <N>ormal functions from the FILE DISPLAY screen in WSHFAX.

Complementing RAM

Palette manipulation is a fast way to complement screen images from the FILE DISPLAY screen. However, we face a more complex situation with the IMAGE DISPLAY screen and the Binary File images in the RAM buffer. When <C>omplement is called from this menu, the contents of the RAM buffer are complemented. This results in a complemented image, no matter what processing options are used. This approach also allows us to save the image in complemented form, something we can't do by just doctoring the display palette. Complementing a byte from RAM is quite simple:

$$bc = 255 - bn$$

where bn is the "normal" value for the byte while bc is the complemented version. When you call <C>omplement from the IMAGE DISPLAY screen, the program steps through each byte of the RAM buffer. It pulls out the value of the byte, subtracting it from 255, and replacing the original with the result. The complemented form of the image appears with all processing functions as well as the <S>ave Image option. To return the image to its original form, one only has to execute <C>omplement again, since complementing a complemented byte returns it to its initial value.

If you would like to experiment with complementation using the BASIC code listing in Figure 10.1, only two lines have to be added. Down in the *sndscrn* subroutine, where the notes indicate the insertion of a video transform routine, insert the following two lines:

$$v = 255 - v$$
$$v = int(v/16)$$

The first line complements the value of v from RAM. The second line does a linear conversion from a range of 0 to 255 for the RAM video to 0 to 15 for display. Whatever image you choose to load will now be a negative version of the image as you would normally view it. If you want to see a closer comparison of the two versions, modify the first line of code as follows:

$$if\ p > 384\ then\ v = 255 - v$$

Now when you display a picture, the left side will have a normal palette while the right side will be complemented.

Complementing the video data can occur anywhere in the sequence prior to display. All you need to know is the maximum video value at that point. For example, the two lines above can be replaced with the following:

$$v = int(v/16)$$
$$v = 15 - v$$

In this case, the first line converts the video to a range of 0 to 15 while the second line complements the result.

While complementing the palette, as we did in the previous section, is fast, it applies to the entire screen. By choosing to complement bytes from RAM or video values prior to display, you can gain much finer control over the process. If needed, you could complement just a specific portion of a line (as we did previously), just certain lines or even specific values for the display video.

FALSE COLOR

I must be up-front and admit that I am not a great fan of false-color display for routine use. To see why, lets display an image in color to see both the advantages and problems presented by false color. First, down in the *sndscrn* subroutine, use the following linear video transform:

$$v = int(v/16)$$

Now, temporarily disable the palette setup by inserting an apostrophe (') in front of each of the lines in the palette routine, starting with $r = 0$ down through *next n*. Now when we go to display, the computer's default color palette will be used and the image will look like a rainbow!

The human eye is very sensitive to color variation. Since the default color palette assigns each of the 16 possible display values a very distinct color value, adjacent pixels that differ by only a single step are clearly differentiated from one another. That's ideal if you want to make fine distinctions between video steps, but it can also create a very chaotic image—especially if there is any noise on the image. It is not uncommon for noise to modulate the video by one grayscale step, but this can rarely be noticed in a complex image using the grayscale palette. As soon as you switch to color, however, those small noise effects are visually amplified and make the display more difficult to interpret.

Figure 10.4—A segment of a NOAA visible-light image, displayed using the 320 × 200 VGA mode with 64 grayscale values. While the spatial resolution of the display is inferior to that obtained with a 640 × 480 display, there is no loss of *image* resolution while achieving a four-fold increase in tonal resolution.

Color has the advantage of effectively amplifying small variations in the image, but it does the same thing for imperfections!

The PALETTE statement in BASIC was defined in the previous section. With four parameters, you can make any step correspond to almost any color:

PALETTE r, ((65536 * blue) + (256 * green) + red)

where *r* is a specific grayscale step (0 to 15), and the variables *blue, green* and *red* are the weights (0 to 63) assigned to the respective primary colors. To define the entire grayscale range, you would need 16 PALETTE statements, one for each value of *r* from 0 to 15. When you have a palette that you like for any specific application, the values could be saved and called from disk files.

In the case of IR images, where we are dealing with a thermal range from hot (black) to cold (white), you might wish to assign a range of color values from red (hot) through oranges, yellows, greens, up through various blues at the cold end of the spectrum. This would provide a very graphic representation of temperature values. For visible light imagery, water features usually come out black (barring sun-glint), so you might want to assign the black end to various shades of blue. Land features occupy the midrange values and could be assigned various shades of green through browns. The upper ⅓ to ¼ of the range could remain in the default grayscale. The result would be the display of visible light pictures where water was blue, land features were green and brown, and clouds

were various shades of light gray through white. It's a cartoon version of what the colors might actually look like if you were up with the spacecraft!

There are more subtle ways to use color that may occasionally prove useful, especially with IR images. Go back to the program in Figure 10.1 and remove the apostrophes (') that disable the palette setup. Now, after the line *next n*, insert the following two lines:

r = 15: blue = 0: green = 0: red = 60
PALETTE r, ((65535 * blue) + (256 * green) + red)

Display the picture and observe the results. Any pixels with the value of 15 will display as a bright red! As written, this pair of lines overrides the palette assignment for video level 15, assigning it a bright red value. If you wanted to identify the most intense centers of a frontal system, this would do the job nicely. Now change *r* in the first line (above) to 7. When you display using this version, red will show up wherever pixel values of 7 appear on the image. In effect, you have identified all the points on the image with an equivalent range of temperatures! Changing *r* will change the video step you highlight, while manipulating values for variable *blue, green* and *red* will change the highlighting color.

BALANCING SPATIAL AND TONAL RESOLUTION

Any program represents a set of compromises, balancing the resources of your computer against various

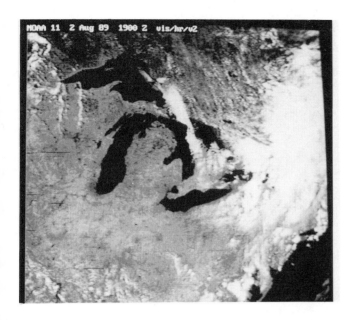

A

B

Figure 10.5—(**A**) A high-resolution subset of a NOAA visible light pass. By sampling a small part of the full-frame image (¼ in this case), the 512 × 480 display mode can be used to display details very close to the theoretical resolving limit for APT imagery. (B) A two-level zoom on the area of northern Lake Michigan. Two small islands, North and South Manitou, can be observed in the lake just west of Grand Traverse Bay. South Manitou is the smaller of the two with a length of just over 5 km. The instantaneous field of view (FOV) of visible light NOAA APT data is 4 km at the nadir. Despite the simplicity of the WSH Interface, there is nothing in the APT (or any other format) that cannot be resolved, subject only to the limitations of the spacecraft sensors themselves.

display objectives. WSHFAX utilizes the 640 × 480 display mode to optimize full-frame display in terms of spatial resolution. At the same time, it provides a range of transform functions to make the most flexible use of the 16-color limit in this mode when using standard VGA displays. If you want to experiment, other trade-offs are quite feasible.

Displaying 64 Grayscale Steps

For example, SCREEN 13 (as opposed to the SCREEN 12 mode we have been using) will provide a 320 × 200 display with a maximum of 256 colors. In grayscale, as opposed to color, this limits us to 64 grayscale values (0 to 63), but that is a four-fold increase in on-screen tonal resolution—at the expense of spatial detail! While a full-frame image wouldn't look very good on a 320 × 200 display, there is no reason why we can't extract a 320 × 200 subset of our image in RAM and display it with 64 grayscale values/pixel. Figure 10.4 shows an example of what such a display can deliver.

Displaying Higher Spatial Resolution

While the most obvious route to increased spatial resolution is an upgrade of your VGA display, there are

plenty of interesting options that will work within the context of the "standard" 640 × 480 option. In writing WSHFAX, I wanted to retain as many image-processing options as possible. I elected to store 8-bit video values in RAM, resulting in a 1:1 correspondence between memory bytes and pixels. In order to fit the various full-frame formats within the limits of the 512 × 480 display "window," the incoming video is sampled prior to storage in RAM and display on the screen.

Let's use WEFAX as an example. The WSH Interface has a clock that will let us take 1024 samples in a WEFAX line, yet we have only 512 pixels available for display. This means that our line sampling will save every-other pixel. Line sampling is a matter of judgment and aesthetics. If we sampled every other WEFAX line, the 800-line image would take up 400 lines of the 480 available for display. The drawback here is that the aspect ratio would be distorted. Instead, I elected to sample two lines and skip one, repeating the sequence until the display screen was filled. This happens after 720 lines have been received, which means the last 80 lines are ignored. The result is very-nearly a perfect aspect ratio, an increase in vertical resolution, and enough of the frame to do a

A

B

Figure 10.6—By varying the rate of sampling, higher resolution display can be achieved if the sampling area is reduced proportionally. (**A**) The default full-screen display of a winter visible light METEOR pass from south to north over the Canadian Maritime Provinces and Quebec. The image has been inverted to place north at the top of the display. Ice-covered James Bay is visible in the upper left while the Gulf of St Lawrence is in the upper right. Lake Manicougan, a meteor-impact feature, can be seen just to the right of the center line in the upper part of the image. The full-frame image was sampled by displaying every fourth pixel on every third line to accommodate the 512 × 480 display area while maintaining a reasonable image aspect ratio.

(**B**) One-quarter of the original image area, centered on the upper right of the original display. This display was obtained by sampling every other pixel on every other line. It shows a significant increase in resolution, compared to the original. The same 512 × 480 VGA display is employed, but it is used to display a smaller portion of the original image.

(**C**) The result of displaying every pixel on every line. Resolution improves, but over a smaller area. This display covers ⅛ of the original image, or ¼ of the display shown in (B). The sampling protocol here introduces some distortion of the aspect ratio, but the increased detail makes the distortion tolerable in most applications.

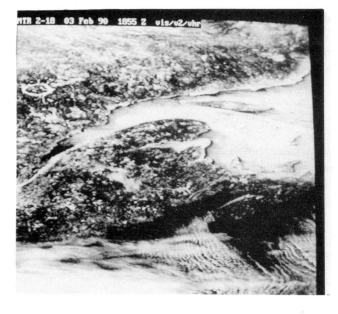

C

creditable job in reconstructing the Earth disk. The solution isn't perfect, but, as noted earlier, almost everything we try to do has trade-offs.

If we accept lower tonal resolution, we can achieve higher spatial resolution with the same computer hardware. This is particularly appropriate for WEFAX. Barring malfunctions, the images are thoroughly processed on the ground. Little can be achieved by further processing. For example, if we do a linear video transform to display *and* store 4-bit pixels instead of 8-bit ones, we can pack two pixels in each byte of RAM storage instead of just one. This effectively *doubles* the number of pixels we can store. We could put this new capacity to use by sampling every line for a total of

A

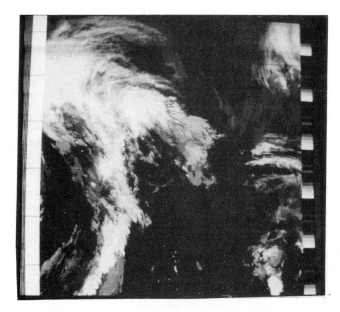

B

Figure 10.7—(**A**) The IR_RAW Binary File, displayed using the WSHFAX <B>inary File Load routine. This represents the raw IR data from a NOAA 12 morning pass. It is typical of TIROS/NOAA infrared data in that most of the picture information is located in the upper half of the display grayscale range. This leads to a light, washed-out image that is lacking in detail. Figure 10.8 shows a histogram plot of the pixel brightness distribution (see RAW plot) for this image. Winter imagery tends to have even less dynamic range than this summer pass, primarily due to low ground temperatures at high latitudes. (**B**) The same image after performing a pixel "slide" (see the RAW + SLIDE histogram in Figure 10.8). The slide has moved all image data down in terms of grayscale so the darkest points of the image are now black. At this point, however, the image still lacks contrast because image data are confined to the lower half of the dynamic range. (**C**) The same image after a pixel "stretch" has been applied (see the SLIDE + STRETCH histogram in Figure 10.8). Image data now occupy the full display dynamic range, showing a major improvement in contrast and observable detail.

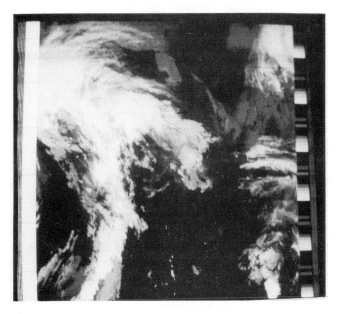

C

682 lines (instead of sampling 480 and skipping 240). Line sampling might then be altered to sample two pixels and skip one, which would use up the remaining RAM capacity. We would have a 768 × 682 image stored in RAM instead of the 512 × 480 image stored by WSHFAX. We wouldn't be able to do any meaningful image processing, but we could pan and zoom from RAM to the 512 × 480 display to take advantage of the additional detail provided by the increased spatial resolution. A similar technique could be used with the various polar-orbit modes. However, you'd probably want to use a nonlinear transform (see later discussions), depending upon what features you wished to emphasize.

There is an alternative strategy that lets us retain all the image processing options, but provides a higher effective display resolution. The idea is to sample a subset of the incoming image at a higher sampling rate. We can use our 512 × 480 display capability to display just part of the image. For example, if we sample every pixel and line for just a portion of each line, we would have a full-screen display of a segment of the image instead of the full-frame. Figure 10.5 shows what can be achieve with such "high-resolution" sampling. In this picture, a segment of a NOAA pass over the Great Lakes, has a *sampling* resolution of 1024 pixels and is essentially equivalent to the maximum resolution the TIROS/NOAA APT format can deliver. We can't display the entire image at that resolution, but we can display the equivalent of about ¼ of the full frame. This is a fundamentally different technique than the <Z>oom function available for File Images in WSHFAX. The <Z>oom routine simply magnifies a portion of the screen display. It makes specific objects bigger, but cannot reveal details that are not already present. Comparing Figure 9.6 in Chapter 9 with 10.6A should make the difference evident. Because of the lower line rate for 120-LPM METEOR images, the 1024-Hz sampling clock permits two levels of high-resolution sampling. Figure 10.7 shows a sequence of METEOR images with progressively higher display resolution.

As noted earlier, many of the spatial and tonal resolution constraints can be overcome by upgrading your VGA display. This typically means a monitor upgrade as well as a new card, but it is now possible to get a 1024 × 768 display with 16, 256, or even more "colors." The only down side to this capability, other than cost, is the fact that the requirements for image file storage go up comparably. WSHFAX stores 512 × 480 8-bit images in 256k Binary Files. If you wanted to do the same thing with 1024 × 768 8-bit files, your files jump to 768k—three times the size required for the 512 × 480 images! Hard-drive capacities seem to increase steadily, while prices per megabyte fall, and the near-future may hold affordable read/write optical storage. Professional users of direct-readout imagery would certainly benefit from the increased resolution afforded by Super-VGA displays, but most users may be content to stick with standard VGA capabilities while watching prices!

EXPANDING IMAGE DISPLAY RANGE

WEFAX, METEOSAT, GMS, and HF WEFAX images are processed prior to transmission and benefit very little from additional processing. In contrast, polar-orbit data (NOAA and METEOR visible and IR) are "raw" and don't always yield the kind of display we might wish for specific applications. The dynamic range of APT modulation for TIROS/NOAA is carefully tailored to be able to handle the maximum range of visible light luminance and IR radiance the spacecraft might encounter anywhere in the world at any season of the year. When our display equipment is

adjusted to optimally display the downlink modulation format, it is typical to encounter a situation where the image data do not actually represent the full dynamic range of the modulation format.

If you boot WSHFAX and load the IR_RAW Binary File, you should see an image essentially identical to that in Figure 10.7A. This is typical of raw (unprocessed) IR data. The image looks very light and washed out. This was a summer pass—winter data are even worse! What's wrong with the picture becomes obvious if we run an analysis of the brightness (actually IR radiance) variations in the image, as shown in the RAW histogram plot in Figure 10.8. There are large numbers of pixels in the upper half of the dynamic range. When the values dip below 90, however, there isn't anything of consequence except for a very small number of pixels, probably derived from grayscale/telemetry information on the side of the image. Fixing an image like Figure 10.7A, can be accomplished in a number of ways. We will tackle the most fundamental approaches first. What we need is a two-step process involving two of the fundamental tools of image processing—the *pixel slide* and the *pixel stretch*.

The Pixel Slide

The purpose of a pixel slide is to set some part of the pixel distribution plot at a definite point in the dynamic range. In this case, we want pixels at the lower end of the IR image data distribution to correspond to black. We want to "slide" the entire pixel distribution over to the left side of the plot—hence the name "pixel slide." For the sake of argument, let's assume that we can ignore all pixel values below 90. We want to use 90 as our baseline and slide the entire distribution down until the pixels that are now 90 will be 0. It is a matter of simple subtraction. For each byte in the RAM buffer we:

1. Recover the byte
2. Subtract 90 from the value
3. Check to see that the result is >0 (to account for the small number of pixels below 90) and if not, set the byte to 0.
4. Replace the pixel in the buffer

If we do so, we end up with a distribution that looks like the RAW + SLIDE plot in Figure 10.8. Note that we haven't changed the plot distribution (the histogram still has essentially the same shape), but all have been moved down by 90. As you can see, there is now a big gap at the upper end of the dynamic range and you might predict that the resulting picture would be too dark. That prediction would be correct. You can see the results of the pixel slide in Figure 10.7B. At first glance, it might seem that we have accomplished very little, exchanging a picture that is too light for one that

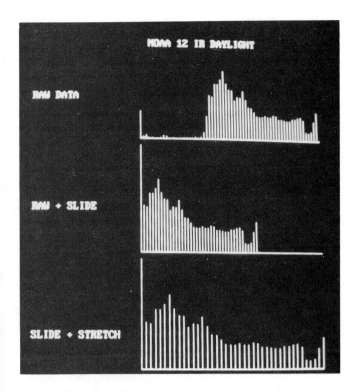

Figure 10.8—Pixel histogram distributions for the images in Figure 10.7A (RAW plot), 10.7B (RAW + SLIDE plot), and 10.7C (SLIDE + STRETCH plot). In all cases, the baseline extends from a pixel value of 0 at the far left (black) to 255 (white) at the far right. The height of each bar in the histogram is proportional to the number of pixels of that brightness/radiance value.

is too dark. Rest assured that this was a necessary first step to perform a first-class magic trick on the next pass.

For the moment, let's digress and try the same thing with our BASIC program (Figure 10.1). Down in the space for the video transform, insert the following three lines of code:

```
v = v − 90
if v < 0 then v = 0
v = int(v/16)
```

The first line offsets each pixel by 90, the second checks to make sure that none of the residual pixels are less than 0, and the last line is our familiar linear transform function. We are not attempting to modify data in RAM with this routine, but doing the slide "on-the-fly." Now execute the program listing using the IR_RAW Binary File. What you should get is something almost identical to Figure 10.7B. You have just performed a basic pixel slide! It's slow and clunky in BASIC, but it got the job done. Shortly we'll see how

to do the job even faster, but, for the moment, the image still needs some fixing.

Pixel Stretch

Prior to the pixel slide, we had an image with a pixel distribution bunched up at the white end of the dynamic range. With the slide, we have a similar bunching at the black end of the range. Fortunately, this is easy to fix using a technique call the "pixel stretch."

The first thing we need to know is the highest value in the pixel frequency distribution. If we can assume that the original maximum pixel value was 255 (a safe bet if you look at the RAW plot in Figure 10.8), and we subtracted 90 from all values when we did the slide, it follows that the highest value in RAW + SLIDE distribution in Figure 10.8 must be 255 − 90 or 165. As we shall see, things become very simple if we can determine what we have to do to the value of 165 to get it back out to 255 (white) where it belongs! Addition (adding 90) is out. If we do that, we simply reverse the effects of the "slide"!

Well, how about multiplication? If we divide 255 by 165, we get a value of 1.54 plus a few decimals. One hundred and sixty-five, multiplied by a correction factor of 1.54, will take the value up to essentially 255. The neat aspect of multiplying by a correction factor is that large pixel values (like 165) are moved quite a bit, but smaller pixel values are displaced less with the same correction factor. At the black end of the scale, 0 doesn't change at all when multiplied by the factor.

If we take such a correction factor (1.54 in this case) and multiply every value in the distribution by the factor, we can expand or "stretch" the distribution so it covers the entire dynamic range. The SLIDE + STRETCH distribution in Figure 10.8 shows the results of applying the correction factor to every discrete video value in the distribution. Note that the darkest areas of the picture are now set to black while the lightest areas are set to white. The shape of the distribution has not been changed. Instead, it has been "stretched" to cover the full dynamic range, hence the term, "pixel stretch." Such a picture should have excellent contrast, and Figure 10.7C shows that this is indeed the case. In addition to the more pleasing range of contrast, all sorts of cloud detail is now visible.

If you want some hands-on experience with our little BASIC module, add the following code to the modified listing that you used for the slide:

```
v = v − 90
if v < 0 then v = 0
v = int(v * 1.54)
if v > 255 then v = 255
v = int(v/16)
```

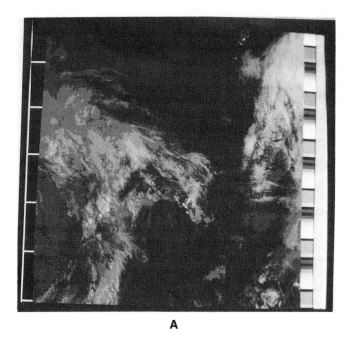

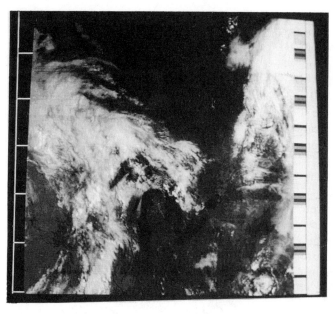

A   B

Figure 10.9—(A) The result of displaying the VIS_RAW Binary File. With proper adjustment of the SAT CONTRAST control and a linear video transfer function, the image appears dark and lacks contrast and detail. (B) The same image after a combination pixel slide/stretch, showing much-improved contrast and considerably more image detail.

The first two lines of the transform are familiar. The first does the slide and the second checks for values below 0. The next two lines are the new ones. The first does the stretch, and the second checks for values greater than 255. Finally, we have the last line, our old reliable linear transform. If you run this code using the IR_RAW Binary File, you should end up with an image that looks virtually identical to Figure 10.7C.

How WSHFAX Does It

At this point you probably won't be surprised to see that WSHFAX will do slides and stretches. What's more, WSHFAX will compute the necessary frequency distributions (they are different for every picture) in addition to doing the actual operations. You will look in vain for "slide" and "stretch" on the program menus. They are hiding under assumed names: *<D>arken* and *<L>ighten*. Both are available from the IMAGE DISPLAY menu.

Boot WSHFAX and use <B>inary File Load to load and display the IR_RAW file. Needless to say, it should look like Figure 10.7A. The function that provides the slide is <D>arken. This is appropriate because that's what it will do to the image. Note that both <D>arken and <L>ighten fall under the Processing Options and that each is marked with an asterisk (*). This is a subtle warning that using them will alter the contents of the RAM buffer. In practice, you would not want to use

either with a *new* image before you had done a <B>inary Save. We don't have to worry about that here, since we are already working with a Binary File.

With IR_RAW displayed, hit key <D> to invoke the <D>arken routine. The screen will clear and a "Scanning image..." notice will be posted. The system is now sweeping through the RAM buffer, tabulating pixel values to construct the frequency distribution—a process that will typically take 5 seconds. Unlike the histograms in Figure 10.8, which are based on *all* the pixels in the image, WSHFAX saves time by sampling every 4th pixel. That may seem like cheating, but you can take my word that over 61,000 samples is a statistically valid sample of a population of almost 250,000!

When the scanning is complete, the screen will clear and you will see a histogram, very much like the RAW plot in Figure 10.8. One difference is that it is up-side down—the baseline is at the top and the bars hang down. Other than that, the shape of the plot will be quite similar to the RAW plot in Figure 10.8. Similar, but *not* identical. (We're sampling, remember?) The image in Figure 10.8 includes *all* pixel values.

At this point, you get to set the black limit for the pixel slide. If you look carefully above the baseline, you'll find a short, vertical pip at the left (black) end. If you press and hold the <R> key, the little pip will slide over toward the right. Pressing <L> will cause it to move left. Use <R> to move it to the right until you have it positioned where a significant number of pixels start to appear. If you overshoot your target, slide back with the <L> key. When the pip is positioned at the point where you want to set your black level, hit <D> for <D>one.

The screen will clear at this point. Within seconds, the slide transform will be executed at all RAM buffer locations and the image will be reposted. This time it should look essentially like Figure 10.7. Obviously, it's too dark, but by now you know that the slide performed by <D>arken is just the first step.

Once you've done the slide, it's time for the stretch—performed with the <L>ighten routine. Key <L> and the screen will clear with the scanning message. When scanning is complete, the frequency distribution of the image will be posted. Surprise! It should look like the RAW + SLIDE in Figure 10.8 (assuming you didn't place the pip at some odd location in the distribution when you used <D>arken).

At this point, you could go through the same drill you did before in setting the black limit. However, this isn't required since we previously did the slide. Just hit <D>. Instead of clearing, the screen header for the histogram will switch to SET WHITE LIMIT, and the little pip will be at the white end (far right) of the baseline. Use <L> to move the pip left to the point that you want to serve as the white limit. Use <R> to move the other way if you overshoot. Press <D> when you get it where you want it. The screen will clear promptly. In just a second or so, the contents of the RAM buffer will be "stretched" and the revised image will be displayed. It should look essentially like Figure 10.7C. You may wish to <B>inary Save this version, so we can work on it later.

Without digital-image processing, it is very difficult to get satisfactory display of IR images. This is especially true if your processing circuits are set up for optimum linearity on both the visible and IR channel. There are processing circuits that can improve on things a bit, but they render the visible channel data unusable if they are left in line. With application of the very simple processing techniques outlined so far, IR images can be just as spectacular as visible light pictures—even in the dead of winter. That's a real benefit because mid-winter sun angles make visible light images mediocre at best.

Once you have mastered the use of <D>arken and <L>ighten on the IR_RAW sample image, do a <B>inary Load on the VIS_RAW image and let's tackle its problems. The picture will look quite dark with a very limited dynamic range (Figure 10.9A). In fact, it looks very much like the IR image after the slide. Fixing it involves exactly the same <L>ighten routine.

When you key <L>, there will be the usual delay while the RAM buffer is scanned and then the pixel histogram will be posted. The SET BLACK LIMIT

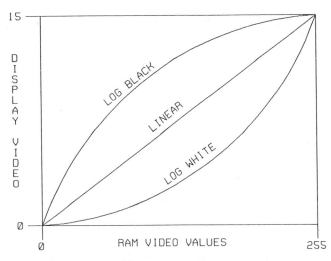

Figure 10.10—Graphic representation of the relationship of RAM video values (0 to 255) and display video values (0 to 15) using a linear and two nonlinear (Log Black and Log White) transfer functions.

option at this point was of no practical use when you did the IR image because you had previously done a slide by using <D>arken. In the case of visible light data, there may be a modest gap between black (far left) and the first significant cluster of pixels. You can move the pip to the right to mark a new black reference, just as you did with <D>arken, hitting <D> when you're done. Then go ahead and mark the white limit, just as you did with the IR sample. At that point the data will be stretched in RAM and the revised image posted to the screen. It should look like Figure 10.9B— a major improvement!

You can, of course, use <D>arken and <L>ighten with images in any mode. In the case of WEFAX, the pictures usually have excellent dynamic range, so little benefit will be realized. The routines are there, however, if poor image setup on the ground delivers a less-than optimum picture. METEOR images typically have a very wide dynamic range and, again, the improvements will not be as dramatic as they are with TIROS/NOAA imagery.

Nonlinear Video Transforms

Up to this point, no matter what we've done to alter the palette or the contents of the RAM buffer, our last step has been to transform the RAM buffer values (with a range of 0 to 255) to display values (range 0 to 15) in a *linear* fashion. Figure 10.10 is a graph of three possible relationships between RAM buffer values and screen display values. The straight-line or linear relationship is the most accurate transform.

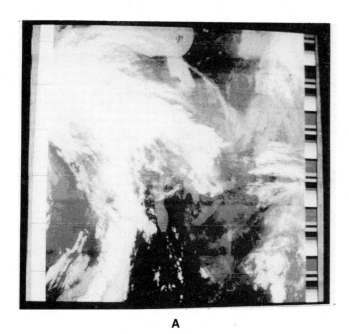

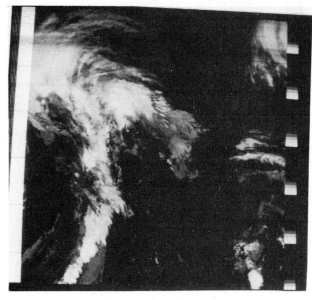

| A | B |

Figure 10.12—(A) The result of applying a Log Black transform to the IR image in Figure 10.7C. As in the case of the visible light example in the previous figure, the result is enhancement of ground detail. (B) The result of applying a Log White video transform to the raw IR data in Figure 10.7A. This provides a quick way to enhance IR image contrast without altering the contents of the RAM buffer.

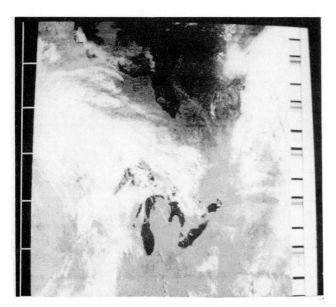

Figure 10.11—The visible light image in Figure 10.9B after a Log Black video transform. This routine can be used to enhance ground detail without altering the contents of the video RAM.

While a straight-line relationship may be "honest," it is far from the only way to make the transformation from RAM to the display screen. Within the limits of the tonal steps available for storage and display, there is almost infinite variation in how we can transfer information from memory to the display screen. The other two curves in Figure 10.10, Log Black and Log White, show two additional possibilities.

You will note, in the case of the Log Black curve, that output video levels rise rapidly at the low end of the range for RAM values, but then taper off. It takes a very large change in the RAM video value to result in a single-step change in the display video. The effect of a curve of this type is to emphasize and enhance the brightness at the low end of the video range while compressing tonal resolution at the white end of the dynamic range. This makes it easier to see things in the darker parts of an image!

WSHFAX has a Log Black function available from the IMAGE DISPLAY menu. Load the processed (contrast-enhanced) version of the VIS Binary File. When the image is passed to the screen from RAM, the default transfer function is linear and you'll have an image similar to Figure 10.9B. Although you can see Michigan and the surrounding lakes, for example, they are really quite dark. To the north, Hudson and James Bay are even harder to see against the dark background. Even though this was a summer pass, it occurred in the morning and sun angles are correspondingly low to the north.

We can reveal more land/water detail by invoking the Log Black transfer curve with the <1> key. When

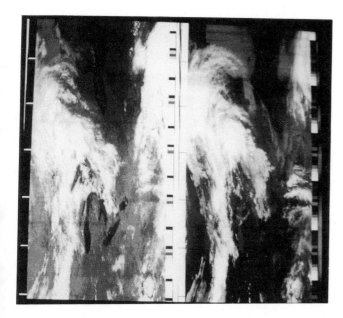

Figure 10.13—Side-by-side display of NOAA raw visible and IR data is rarely satisfactory since there is a pronounced difference in dynamic range for the two formats. Visible data typically looks too dark while IR data is too light (see Figure 9.11 in Chapter 9). By applying differential enhancement routines to the two halves of the image, display dynamic range can be optimized for each format, making them much more comparable. This is the same image shown in Figure 9.11, but displayed with differential processing of the visible (left) and IR (right) image data.

you do so, the image will paint down again. You'll see a major improvement in your ability to see land/water detail, at the expense of compression of the white cloud data. Your display should look something like Figure 10.11. Hit key <2> for the Linear transform and the original version of the image will scroll down. You might also load the processed version of the IR image (which should look like Figure 10.7C). Press key <1> for the Log Black transform and an image similar to Figure 10.12A should appear. Now the ground detail is quite evident and you should be able to see temperature gradients in Lake Michigan. Put simply, Log Black enhances detail at the black end of the dynamic range.

Although the default transform is Linear, once you select a function like Log Black it will remain the operating transform until you select <2> for Linear or begin to display an image. The Linear transform is always used when acquiring an image.

Curves similar to the Log White example in Figure 10.10 behave quite differently from Log Black. They compress data at the dark end of the dynamic range and expand the detail at the white end of the dynamic range. This makes them particularly suitable for quick enhancements of raw NOAA IR data. Load the IR_RAW Binary File. With the Linear transform, you will get an image identical to Figure 10.7A. Now key <3> to do the Log White transform. You should get something similar to Figure 10.12B. There is a significant improvement in cloud detail. The image is not as good as the fully processed version (Figure 10.7C), but it is better than the original and the routine is very fast.

Log Black, Linear, and Log White are all variable transfer functions. Unlike <D>arken and <L>ighten, they do not modify the RAM buffer data. Basically, they reflect variations in values loaded to the 256-entry look-up table that the program uses to do RAM to screen transfers. A table-type system is very flexible, although the code for screen dumps has to be changed. To change the transfer function, you simply alter the entries in the look-up table.

If you are writing your own code, it is fairly easy to insert routines that allow the construction of a number of custom transform curves that can be saved as disk files. To invoke one of them, load the specific file and POKE the entries into the table prior to dumping the data to the screen from RAM.

CONCLUSION

As you can see, the options for digital image manipulation are open-ended. We have touched on some of the high points in this chapter, but these topics are only the tip of a very fascinating iceberg. If you like to tinker, the very simple BASIC program we have been using will let you manipulate WSHFAX Binary Files. It is slow, but using the program requires no great programming experience. Feel free to experiment—you can't hurt anything! You can also elaborate the program and insert additional features, such as saving the altered images you create.

Glossary

I t is impractical for me to hope to define every term with which you may not be familiar. This glossary includes those terms that are most relevant to satellite-station operations, but largely excludes terms used in general electronics and radio technology—including most antenna terms. A good general reference covering most of these areas is *The ARRL Handbook*, published annually by the American Radio Relay League.

Active filter—an amplifier (usually an op amp IC), used in conjunction with specific resistor and capacitor values to create a filter with desired characteristics (see Filter).

Address—combination of logic values on an address bus that specifies the location of a particular byte of data in a memory chip.

Address bus—the combination of individual address lines in a microprocessor system that, in conjunction with other signals, defines memory locations, I/O ports, or other elements in the memory map.

AM—*Amplitude Modulation*. A method of placing information on an RF or audio-frequency carrier by varying the amplitude of the modulating frequency.

Amplifier—an electronic circuit that provides gain.

Amplitude—the extent to which the voltage or current of an RF or audio signal varies compared to zero or a mean level.

Analog signal—an electronic signal that is represented by varying levels over a continuous range rather than in discrete steps.

Anomalistic period—a value for the period of an orbiting spacecraft determined by dividing the total number of minutes per day (1440) by the number of orbits per day.

Apogee—the high point (greatest distance from the surface) or a satellite orbit. (see Eccentricity and Perigee).

APT—*Automatic Picture Transmission*. An acronym describing the automatic transmission of images by polar-orbiting spacecraft.

Ascending pass—a pass originating to the south of a ground station and ending to the north of it. Ascending passes produce inverted images on display systems that don't account for the directional difference between a north-south and south-north pass.

Assembler—a computer program using mnemonic commands that permits the programmer to create, in one or more steps, a machine-language program for a specific microprocessor.

Azimuth—the compass direction (usually expressed in true degrees) required to point a ground station antenna at a distant spacecraft.

Az-el—an antenna mount that provides independent control of the azimuth and elevation of an antenna.

Bandpass filter—see Filter.

Bandwidth—the range of frequencies passed by a filter circuit. Since filters are never perfect, bandwidth is usually expressed as the total frequency range producing a 3-dB decrease in signal level.

BASIC—*Beginners All-purpose Symbolic Instruction Code*. The most widespread of the high-level programming languages for microcomputers that uses English-like statements and commands to create a computer program. There are generally two kinds of BASIC operating systems. In the most common form, interpreted BASIC, the BASIC statements or commands are converted to machine-language instructions at the time the statement or command is executed. This is the slowest form of BASIC. In *compiled* BASIC systems (such as QuickBASIC), a separate *compiler* program converts the BASIC commands and statements into machine-language code that represents the actual computer program. Compiled BASIC programs are notably faster in execution than interpreted programs because runtime conversion of the statements and commands is unneccessary.

Beamwidth—the width of an antenna pattern, usually expressed in degrees, that results in a 3-dB decrease in signal level at the beamwidth pattern limits.

Bearing—the combination of elevation and azimuth required to point an antenna at a specific spacecraft at any given moment.

Bidirectional—a data port through which data can travel both to and from a specific device.

Binary—a system of expressing numerical values to the base 2.

Bit—the smallest unit of data in a computer system. Any specific data bit can either be high (1) or low (0).

CDA—*Control and Data Acquisition*; acronym for the ground control stations for the TIROS/NOAA polar orbiters.

Chip—vernacular term for an integrated circuit.

COSMOS—a generic name applied to a wide range of Russian research satellites. Some COSMOS spacecraft transmit very-high-quality 240-LPM meteorological images.

CRT—*Cathode Ray Tube*; a television-like image display device.

Crystal—one or more thin plates of quartz, mounted in a hermetically sealed enclosure. Crystals are used to control the frequency of oscillator circuits and are used as passive filter elements in receiver IF circuits.

Decay—the gradual decrease in orbital period and spacecraft orbital altitude caused by residual atmospheric drag.

Descending pass—a pass that originates to the north of a ground station and terminates to the south of it.

Downconverter—a series of modules, usually consisting of one or more RF preamplifiers, a local oscillator, mixer, and IF amplifier designed to convert S-band (1691-MHz) signals to the 136 to 138-MHz VHF satellite band.

Eccentricity—the ratio of apogee and perigee. A circular orbit has an eccentricity of 1.

Elevation—the bearing in degrees, relative to the horizon, of the principal lobe of an antenna. At an elevation of 0° the antenna is aligned with the horizon. At an elevation of 90° it is pointing straight up.

ERP—*Effective Radiated Power*. The effective power output of a transmitting system (usually expressed in watts or milliwatts (thousandths of a watt), taking into consideration the power output of the transmitter, losses in the transmission line between the transmitter and the antenna, and the antenna gain and pattern characteristics.

ESRO—*European Space Research Organization*. A consortium of European nations operating specific space-research projects such as the METEOSAT geostationary satellite system.

Facsimile—a system for reproducing a satellite (or other) image directly onto photographic or other specialized paper media or a computer display.

Fax—another word for facsimile.

Feed horn—one of the simplest systems for intercepting the RF energy reflected by a parabolic antenna and transferring that energy to a transmission line.

Filter—an electronic circuit composed of either passive (crystals, ceramic filters, coils and capacitors) or active devices (usually operational amplifiers) designed to pass a certain range of frequencies while providing some measure of reduction in signals at other frequencies. *Band-pass* filters (such as the crystal and ceramic filters in receiver IF circuits), or the active filter at the input of the video circuit of Chapter 5, are designed to pass a specific range of frequencies (defined by the design center frequency and bandwidth of the filter), while rejecting signals above and below those frequencies. *Low-pass* filters, such as those in the post-detector video circuits, are designed to pass signals below a specific frequency, rejecting signals above that frequency. *High-pass* filters reject signals below a specific frequency.

Focal length—the distance from the center of the surface of a parabolic antenna to the point at which the RF energy is brought to a focus (the focal point).

Geometric distortion—foreshortening of a view toward the horizon resulting from the wide field of view of the image-scanning radiometers. This distortion was obvious on the ITOS/NOAA spacecraft (NOAA 2-5), but is essentially removed by on-board computer processing of the HRPT data stream in the TIROS/NOAA (NOAA 6-11) spacecraft. (Only NOAA-9, -10 and -11 are presently active.)

Geostationary orbit—an orbit (approximately 22,000 miles high) the plane of which lies on the equator. Such an orbit has a period of 24 hours (1440 minutes). If the direction of movement of the spacecraft in this orbit is the same as the direction of the rotation of the earth below, the spacecraft subpoint on the equator will not change.

GIF—*Graphical Interchange Format*. A set of graphics standards developed by CompuServe that permits grayscale images to be displayed on a wide range of computers with different graphics standards.

GMS—*Geostationary Meteorological Satellite*. A spacecraft, similar to GOES, which is operated by the Japanese government over the western Pacific.

GMT—an obsolete term; see UTC.

GOES—*Geostationary Operational Environmental Satellite*. The geostationary meteorological satellite system, consisting nominally of three GOES spacecraft, operated by the US Government.

Ground track—projection on a map of successive spacecraft sub-point positions (see Track).

Horn—see Feed horn.

HRPT—*High Resolution Picture Transmission*; acronym for the high-resolution imaging system employed by the TIROS/NOAA spacecraft. Medium-resolution APT data are derived by sampling this data stream using the on-board computer.

Inclination—the angle of the orbital plane of a spacecraft relative to the earth's equatorial plane.

Infrared—(IR) electromagnetic energy with wavelengths somewhat longer than that of the visible-light spectrum. IR energy is essentially various wavelength of heat radiation lying beyond the red end of the spectrum.

Injection—the process of insertion of a satellite into orbit at a precise angle and speed to achieve the desired orbital geometry.

Interlaced—a system by which a complete image is built up on a CRT by scanning one field consisting of even-numbered lines followed by a second field of odd-numbered lines. Such a system is used in broadcast TV to maximize display resolution while minimizing flicker. Each field requires 1/60th of a second for display; the entire interlaced picture (a *frame* consisting of two fields) is produced in 1/30th of a second.

Intermediate Frequency (IF)—a frequency to which an RF signal is converted (usually lower than the signal frequency in most simple receivers) where it is more convenient to obtain the required signal gain while filtering the signal to obtain a specified bandwidth.

IR—see infrared.

Keplerian elements—a moderately complex set of numerical values that completely describes the nature of the orbit of a spacecraft. Keplerian elements for most satellites are routinely compiled by NASA and are available through a number of satellite bulletin board systems. Sophisticated tracking programs use such elements to gain the required precision for accurate tracking.

LPM—*L*ines *P*er *M*inute; defines the horizontal scanning rate of the spacecraft imaging system that must be duplicated for proper image display on the ground-station equipment.

Meteor—the primary polar-orbiting spacecraft series operated by the Russians. Meteor spacecraft have slightly different orbital characteristics, depending upon the series, but all provide 120-LPM visible-light imagery. The latest spacecraft of this series have been testing IR imaging systems.

METEOSAT—the geostationary spacecraft operated by the European Space Research Organization.

Mosaic—A composite image created by combining individual display segments to produce an image with greater geographical coverage. The cover photo is a mosaic consisting of four segments: the northern and southern segments of an overhead and western pass.

Multi-spectral—an imaging system with individual sensors that respond to different portions of the spectrum. Current TIROS/NOAA spacecraft have a 5-channel instrument with one visible-light sensor and four IR sensors.

NOAA—*N*ational *O*ceanic and *A*tmospheric *A*dministration; spacecraft: the designation of a TIROS spacecraft after launch; agency: NOAA is the agency of the US Department of Commerce responsible for US governmental meteorological services.

Nodal period—the period, in minutes, between successive north-bound equatorial crossings of a spacecraft.

Noise—essentially random RF energy originating in signal processing circuits or from external man-made or astronomical sources.

Non-interlaced—a display system where the final image is created by a single scan of the CRT display tube.

Path loss—the difference between the effective radiated power of an RF source and the power available at the receiver site. Path loss is related to the distance between the source and receiver and the operating frequency.

Perigee—the low point in a satellite orbit. (See Apogee and Eccentricity).

Period—a general term for the time required for one orbit of the earth (see Anomalistic Period and Nodal Period).

Phasing—the processing of getting the display system in step with the image transmitter so that the start of image lines from the transmitter corresponds to the start of a line on the display. An out-of-phase image has the start of one image line somewhere in the main image area rather than at the edge (usually left) of the image.

PEL—*P*icture *El*ement (see pixel).

Pixel—the individual component of an image scan line defined by a single video sample in a digital image system. The greater the number of pixels in a line, the greater the display spatial resolution.

Polar mount—a mounting system for geostationary antennas where the main axis of the system is oriented parallel to the earth's polar axis. This permits an antenna to sweep the entire geostationary arc (Clarke Belt) using a single motorized drive, as opposed to individual control of azimuth and elevation.

Polar orbit—strictly speaking, an orbit whose plane intersects the poles of the earth. In practice, the term is loosely applied to the near-polar orbits of the TIROS/NOAA and Meteor/COSMOS spacecraft.

Probe—the RF pick-up element in an S-band feed horn.

Programmable scanner—a scanning receiver whose operating frequencies are controlled by a frequency synthesizer set by front-panel switches.

Quadrant (quads)—segments of the earth disk transmitted via WEFAX through the GOES geostationary spacecraft.

Rotator—a motorized system for remotely controlling the azimuth or elevation of an antenna system.

S-band—a relatively low-frequency microwave band including the 1691 and 1694.5-MHz frequencies used by geostationary weather satellites.

S/N—*Signal* to *Noise* ratio; the relationship of the power of the desired signal relative to total received power, including undesired noise. The higher the signal to noise ratio, the better the quality of the received signal.

Squelch—an electronic system for cutting off the audio output of a receiver when no signal is present.

Subcarrier—an audio tone (typically 2400 Hz) used to convey the amplitude-modulated video data.

Subpoint—The point on the earth's surface immediately below a spacecraft at any point in time.

Sun synchronous—an orbit in which the orbital plane precesses at slightly less than one degree per day, resulting in a situation in which the spacecraft passes overhead at essentially the same solar time throughout all seasons of the year.

Synchronization—the process of matching line scanning rate in the ground-station display equipment with that of the image transmitting system. Very slight synchronization errors result in tilted or skewed images, while larger errors render the image unrecognizable.

Synchronous motor—a motor designed to maintain an operating speed that tracks the frequency of the ac power signal. Such motors, controlled by a crystal or tuning-fork time standard, are used to maintain synchronization in electromechanical fax recorders.

Synthesizer—a complex of digital circuits that create or synthesize an RF or audio signal at some specific frequency.

TIROS—*T*elevision *I*nfra*R*ed *O*perational *S*atellite; the pre-launch or series designation for the current US polar-orbiting weather satellites. The acronym is the same one used for the very first operational weather satellites. (It is a bit of an anachronism in that none of the current spacecraft employ the television vidicon tubes of their predecessors.)

TNL—*T*hermal *N*oise *L*evel; the internal noise power level (usually expressed in dBmw [decibels per milliwatt]) of a receiving system. TNL is largely determined by the gain and noise figure characteristics of the early RF amplifiers in a receiving system. The TNL should be as low as possible because the desired signal must exceed TNL by 10-20 dB to achieve noise-free images.

Track—the process of adjusting the elevation and azimuth of an antenna to follow a spacecraft for the duration of a pass.

UTC—the correct term to use for time referenced to the prime meridian. Other commonly used terms are GMT (obsolete) or Z (Zulu) time. All spacecraft data are disseminated in terms of UTC time and date, but labeled GMT.

VAS—*V*ertical *A*tmospheric *S*ounder; the high-resolution, multi-spectral imaging system employed by the GOES spacecraft.

VISSR—*V*ery high-resolution *I*nfrared *S*pin *S*can *Ra*diometer; the GOES imaging instrument.

WEFAX—Weather facsimile; an acronym (*WE*ather *FA*X) for the medium-resolution imaging service, including pictures and charts, provided by the various geostationary weather satellites.

Appendix

Parts and Equipment Suppliers

Unlike the situation a few years ago, there is now a comparatively large number of companies servicing the needs of weather-satellite experimenters. Throughout this edition, I have noted specific manufacturers and products. Some were discussed at greater length than others. This does not necessarily represent my assessment of the relative worth or value of the products—it simply reflects my familarity with the equipment line or product on the basis of my own experience. Because I cannot hope to sample everything, my experience is not all-inclusive. Similarly, the listing that follows will not be complete since I cannot hope to be familiar with all of the possible vendors and products. Old products will be upgraded or replaced, and new models and companies enter the field.

Computer bulletin boards, JESAUG, and reviews and advertisements in Amateur Radio journals such as *QST*, *73 Magazine*, and *CQ* (and their peer publications in other areas of the world), can help to keep you up to date on new product lines. By all means, write for literature and seek out photographic repro-

ductions of typical products. Remember that systems designed for HF facsimile use may *not* be compatible with the AM format used by weather satellites. Also, be aware that a system may produce reasonable output on weather charts, but may be deficient in handling grayscale imagery. The examples included throughout this edition represent the kind of quality you can expect with a very modest investment in your station.

The following key will be used in the *Products* line following each vendor:

P—General electronics parts
S—Scan-converter hardware/software
SA—S-band antenna systems
SC—S-band converters
SP—S-band preamplifiers
VA—VHF antenna systems
VP—VHF preamplifiers
VR—VHF receivers
X—Crystals

VENDOR LISTING

Advanced Receiver Research
 PO Box 1242
 Burlington, CT 06013
 tel 860-582-9409
 Products: VP

Amateur Electronic Supply
 5710 West Good Hope Road
 Milwaukee, WI 53223
 tel 800-558-0411
 Products: VR

Cushcraft Corporation
 48 Perimeter Road
 Manchester, NH 03108
 tel 603-627-7877
 fax 603-627-1764
 Products: VA

DARTCOM
 Postbridge
 Yelverton
 Devon PL20 6SY
 UK
 Products: VR

Down East Microwave
954 Rte 519
Frenchtown, NJ 08825
tel 908-996-3584
fax 908-946-3072
Products: SP

Greg Ehrler
105-53 87th Street
Ozone Park, NY 11417
Products: SP

Hamtronics Inc
65-Q Moule Road
Hilton, NY 14468
tel 716-392-9430
fax 716-392-9420
Products: VP, VR (kits and wired and tested)

International Crystal Manufacturing
10 N Lee
Oklahoma City, OK 73126-0330
tel 800-725-1426 or 405-236-3741
fax 800-322-9426
Products: X

Jameco Electronics
1355 Shoreway Road
Belmont, CA 94002
tel 800-831-4242
fax 800-237-6948
Products: P

JDR Microdevices
1224 South 10th Street
San Jose, CA 95122-4108
tel 800-538-5000 (orders)
    408-494-1400
    408-494-1430
fax 800-538-5005
Products: P

Metstat Products, Inc.
1257 Genmeadow Lane
East Lansing, MI 48823
tel 517-773-0667
Products: S

Quorum Communications, Inc.
8304 Esters Blvd
Suite 850
Irving, TX 75063
tel 214-915-0256
bbs 214-915-0346
fax 214-915-0270
E-Mail: info@qcom.com
Products: S, SC, VR

Software Systems Consulting
615 South El Camino Real
San Clemente, CA 92672
tel 714-498-5784
fax 714-498-0568
Products: S

Spectrum International
PO Box 1084
Concord, MA 01742
tel 508-263-2145
fax 508-263-7008
Products: SA, SC, SP, VA, VP

Vanguard Labs
196-23 Jamaica Avenue
Hollis, NY 11423
tel 718-468-2720
Products: VA, VP, VR

Wilmanco
5350 Kazuko Court
Moore Park, CA 93021
tel 805-523-2390
Products: SC, SP

Index

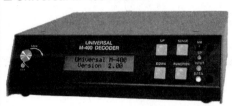

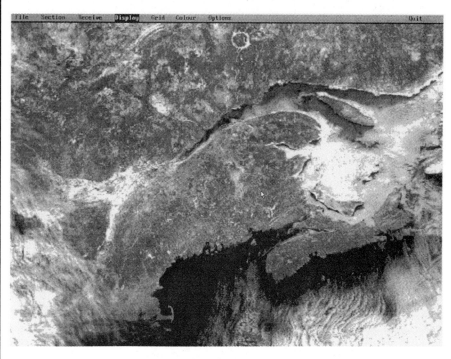

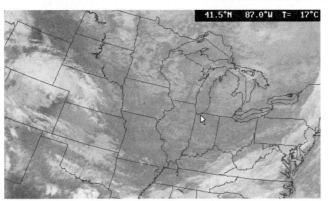

Timestep

North America from Meteosat 3

Whole Earth visible image

Infrared image of North America with temperature readout

PDUS

Timestep

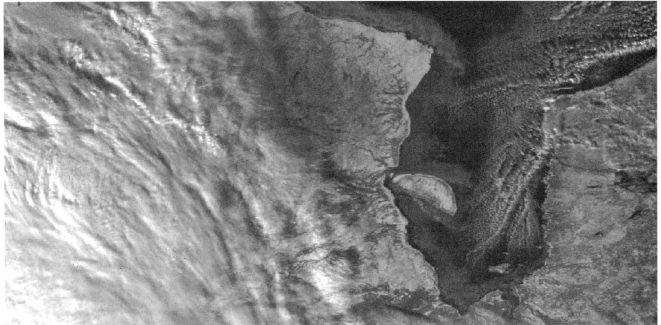
HRPT Visible image of the southern end of Hudson Bay

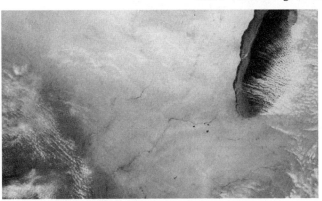
Heat Emissions from Power Stations in Illinois

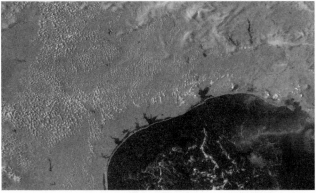
Texas Coast

HRPT

Advanced Very High Resolution Radiometer - High Resolution Picture Transmission - the titles say it all! Quite simply, HRPT is the ultimate in remote sensing; no other system comes close. Now, incredibly, HRPT is easily within reach of any amateur. Stunning images with 1.1km ground resolution are available in your home, classroom or university. Rivers, towns, cities, mountain ranges, lakes and even power stations can be seen with this system.

10 bit images in 5 spectral bands are available. In one single pass, over 80MB of data can be collected. The user friendly, mouse driven HRPT software has many features such as: latitude and longitude gridding, political boundaries, real time pan and zoom, contrast stretch, temperature readout and more. All of the data transmitted is stored for analysis by other platforms or systems if required.

The NOAA AVHRR sensors transmit 10 bit data. Software that only receives 8 bits throws away a lot of information. 8 bit software resolves 256 intensity levels; 10 bit software resolves 1024 levels. Since the satellites rarely achieve their full dynamic range this ability is very important. Automatic temperature readout on all infrared bands is as easy as moving the mouse pointer. Temperature resolution is of course enhanced by having 10 bit data.

The software will automatically recover from any signal failure or dish mistracking. This means you always get a usable image.

Systems can be configured with a 3 foot dish, are FCC Part 15 approved and have ISO9002 quality accreditation. Call for a full colour brochure.

Timestep PO Box 2001 Newmarket CB8 8XB England   Tel. 0440 820040 Fax. 0440 820281

Distributed in the USA by : Spectrum International Inc. PO Box 1084 Concord Mass. 01742 Tel. (508) 263-2145 Fax. (508) 263-7008

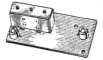
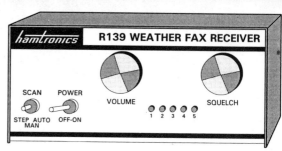

MultiFAX SOFTWARE

Software is used to control the **MultiFAX** demodulator as well as capture and process (enhance, zoom, grid, colorize, etc.) the resulting satellite images. This software must be fast, powerful, and easy to use. MFMAP version 7 software (MFMAP7) sets new standards in all these areas:

✓ To provide speed, much of MFMAP7 is written in assembly language. Speed is one of the reasons the software can support writing the incoming image directly to disk (or RAM disk), display the image on the screen, and track the source satellite - all at the same time!

✓ Review the new features and complete software feature summary listed below to get a feel for the power of the MFMAP7 software. You won't find **any other** image capture package that provides this much function with FULL satellite resolution at such a low price!

✓ Finally, MFMAP7 is easy to use. A "graphical user interface" with mouse support makes the software very easy to learn and use with simple, uncluttered screens and menus. A complete, clear manual is also included.

➤ MFMAP7 can output the **elevation and azimuth** of the polar orbiting satellites right on the record screen, <u>real-time</u>, as you're capturing and displaying the image! Perfect for users that track satellites with manually controlled rotors!

MFMAP7 Setup Screen

➤ For users with the **Kansas City Tracker:** MFMAP7 now provides automatic, computer controlled, tracking capability while recording the polar orbiting weather satellites. No need for two separate computers to record and track! MFMAP7 will send the satellite location information to your KCT driver directly *during the record function.*

New Software Features

➤ MFMAP7 now has **satellite tracking built in!** No need to purchase standalone satellite tracking software at prices higher than the complete MFMAP7 upgrade price! View the satellite track and visibility "footprint" on a world map with "fast forward" and pause capability, view the orbital elements, and print a schedule for the satellite of interest to the screen or printer!

Frame Cursor for Zooming

➤ The **Record Level Meter** has been changed to a graphical display of signal level as the satellite signal is sampled. The new display works like an oscilloscope displaying signal level and holding the display until a new line of data is sampled. You can achieve perfect level settings the first time and every time. Fine tune recordings to optimize the visual or IR image from NOAA satellites.

➤ A new **Record Option for the NOAA** satellites using the demodulator onboard crystal clock is now available. This new feature will prevent image "breakage" during severe signal fades when recording under less than ideal conditions.

➤ **Image enhancement** is easier than ever with the new **Palette Function**. The cursor (arrow) keys can be used to vary brightness and contrast of your image over the complete range of gray scale. Previous methods, including the popular programmable function keys, are still available for custom instant image enhancement.

➤ New enhancements to **MFREC** and **TIMER** software make unattended recording a snap. MFREC is a <u>completely</u> command line driven record program than now includes *AutoStart* for NOAA, Meteor, GOES, and Meteosat images. The *Show* function is also available for GOES and Meteosat. Given a few command line parameters, MFREC can be "called" from any customized software or batch file and capture an image from any WEFAX source that MFMAP7 supports. TIMER is a "Front End" calling program for MFREC that allows up to 200 timed record events to be programmed from a simple schedule file.

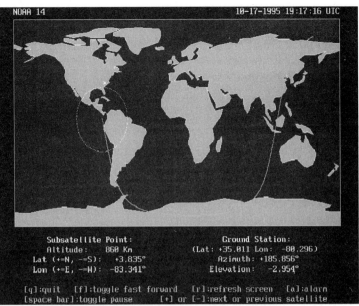

Satellite Tracking on World Map

Minimum requirements: IBM Compatible Computer with 640 KB Memory and either 1) An 8 or 16 bit ISA slot for the internal card **OR** 2) A parallel port (LPT1, LPT2, or LPT3) for the external unit • VGA Card and Monitor • Hard Drive or RAM Disk with 4MB Available Space • Receiver and Simple Antenna (dish not required for high resolution polar orbiting satellites)

Software Summary

Version 7 Software Features Include: Integrated Satellite Tracking • Kansas City Tracker Support • Capture Images to Hard Disk or Memory • PLL Sampling • NOAA, Meteor, GOES, Meteosat, HF Fax • "Point & Shoot" User Interface with Mouse Support • 1024x768x256 Colors/64 Gray Levels • Zoom • Simple, Powerful Image Enhancement • 10 User Definable Enhancement Palettes • False Colorization • Unattended Recording • Visible and IR • Animation • Calibrated IR Temperature Readout • "3D" Enhancement • Use Your Images with Hundreds of Other Programs • Printer Support • 2-3 Mile Resolution (NOAA) • 3.5 Million 8 Bit Pixels for full NOAA Recording • Latitude/Longitude and Map Overlay (US included) • Reference Audio Tape • Clear, Complete 85+ page Illustrated User's Manual • Much More...

Please write or call for more details on MultiFAX Demodulators, receivers, antennas and the MultiFAX GOES/Meteosat 1691-137.5 MHz downconverter. Internal demodulator with software: Just $289 plus S&H.

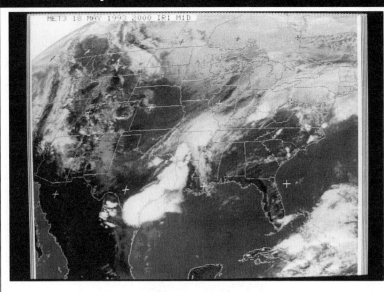
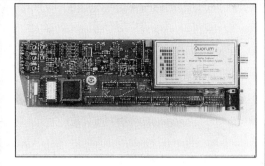

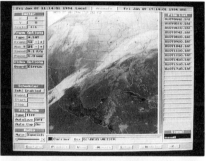

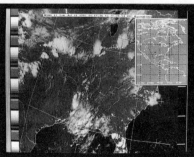

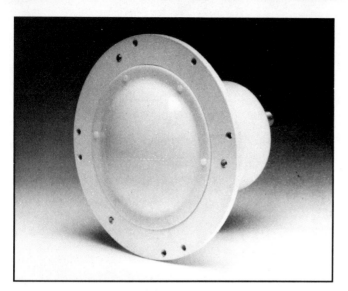
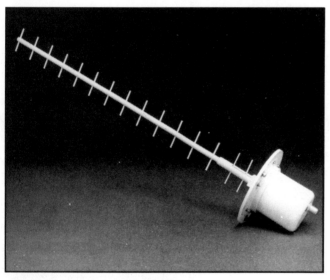

Your **HAM RADIO LICENSE** is only hours away with **THE ARRL TECHNICIAN CLASS VIDEO COURSE**

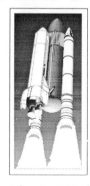

About the American Radio Relay League

The seed for Amateur Radio was planted in the 1890s, when Guglielmo Marconi began his experiments in wireless telegraphy. Soon he was joined by dozens, then hundreds, of others who were enthusiastic about sending and receiving messages through the air—some with a commercial interest, but others solely out of a love for this new communications medium. The United States government began licensing Amateur Radio operators in 1912.

By 1914, there were thousands of Amateur Radio operators—hams—in the United States. Hiram Percy Maxim, a leading Hartford, Connecticut, inventor and industrialist saw the need for an organization to band together this fledgling group of radio experimenters. In May 1914 he founded the American Radio Relay League (ARRL) to meet that need.

Today ARRL, with more than 170,000 members, is the largest organization of radio amateurs in the United States. The League is a not-for-profit organization that:

- promotes interest in Amateur Radio communications and experimentation
- represents US radio amateurs in legislative matters, and
- maintains fraternalism and a high standard of conduct among Amateur Radio operators.

At League headquarters in the Hartford suburb of Newington, the staff helps serve the needs of members. ARRL is also International Secretariat for the International Amateur Radio Union, which is made up of similar societies in more than 100 countries around the world.

ARRL publishes the monthly journal *QST*, as well as newsletters and many publications covering all aspects of Amateur Radio. Its headquarters station, W1AW, transmits bulletins of interest to radio amateurs and Morse code practice sessions.

The League also coordinates an extensive field organization, which includes volunteers who provide technical information for radio amateurs and public-service activities. ARRL also represents US amateurs with the Federal Communications Commission and other government agencies in the US and abroad.

Membership in ARRL means much more than receiving *QST* each month. In addition to the services already described, ARRL offers membership services on a personal level, such as the ARRL Volunteer Examiner Coordinator Program and a QSL bureau.

Full ARRL membership (available only to licensed radio amateurs) gives you a voice in how the affairs of the organization are governed. League policy is set by a Board of Directors (one from each of 15 Divisions). Each year, half of the ARRL Board of Directors stands for election by the full members they represent. The day-to-day operation of ARRL HQ is managed by an Executive Vice President and a Chief Financial Officer.

No matter what aspect of Amateur Radio attracts you, ARRL membership is relevant and important. There would be no Amateur Radio as we know it today were it not for the ARRL. We would be happy to welcome you as a member! (An Amateur Radio license is not required for Associate Membership.) For more information about ARRL and answers to any questions you may have about Amateur Radio, write or call:

ARRL Educational Activities Dept
225 Main Street
Newington CT 06111-1494
(860) 594-0200
Prospective new amateurs call:
800-32-NEW HAM (800-326-3942)

Notes

Notes

Notes

FEEDBACK

Please use this form to give us your comments on this book and what you'd like to see in future editions.

Where did you purchase this book? □ From ARRL directly □ From an ARRL dealer

Is there a dealer who carries
ARRL publications within: □ 5 miles □ 15 miles □ 30 miles of your location? □ Not sure.

License class:

□ Novice □ Technician □ Technician with HF privileges □ General □ Advanced □ Extra

Name _____

_____ Call sign _____

Address _____

City, State/Province, ZIP/Postal Code _____

Daytime Phone (___) _____ Age _____

If licensed, how long? _____ ARRL member? □ Yes □ No

Other hobbies _____

_____

Occupation _____

From _____

_____

_____

EDITOR, WEATHER SATELLITE HANDBOOK
AMERICAN RADIO RELAY LEAGUE
225 MAIN ST
NEWINGTON CT 06111-1494

please fold and tape